Charles Fort is the controversial author whose books—*New Lands, Prophet of the Unexplained, Lo!* and *The Book of the Damned*—have challenged the blind dogmas of science in their examination of mysterious and startling occult and psychic phenomena.

In this, his classic work, he records over 1,000 extraordinary phenomena for which there is no "scientific" explanation, defying our conventional way of looking at things and presenting a startling new picture of what is "rational" and what is not.

The Book of the Damned is much more than a compendium of weird facts and events—it is, quite simply, one of the most important and revolutionary books ever written.

D1300326

THE BOOK OF THE DAMNED

Charles Fort

SPHERE BOOKS LIMITED
30/32 Gray's Inn Road, London WC1X 8JL

First published in Great Britain by Sphere Books Ltd under the
ABACUS imprint 1973
Copyright © Boni and Liveright 1919
Copyright © Holt, Rinehart and Winston Inc 1941
Reprinted 1974 (in ABACUS)
Sphere edition published 1979

Published by arrangement with Mrs. Tiffany Thayer

Set in V-I-P Times
Printed in Great Britain by
Cox & Wyman Ltd,
London, Reading and Fakenham

Preface

by

DONALD A. WOLLHEIM

Had you been standing in a certain crater valley on the moon a few months ago you would have suddenly experienced what people have come to call a "Fortean" phenomenon. For a few seconds, the crater would have been bathed in an inexplicable reddish glow, lighting up the area from no apparent source in a faint but definitely real, eerie blood-coloured glow.

And had you told the others with you what you had seen they would doubtless have scoffed, pointed out that there was no possible source for an illumination on the moon—there had been no eruption, no moon-being had turned on a light, nothing had exploded in the sky, and so forth. In short, they would likely have disbelieved you, and filed your observation in their own book of material damned by science for its improbability.

Nevertheless you would have been right and the scoffers wrong. For the reddish glow on that lunar scene had actually occurred, and it was not subject to normal lunarian scientific explanation. The origin was a new type of light beam invented on another planet—the third from the sun—and that special type of light projector had been turned on the moon from that other world for a few seconds.

To the moon-being, this light baffled respectable conservative scientific analysis; therefore, it would have been best to call it an illusion or just ignore it.

Reading *The Book of the Damned* over forty years after it was first compiled, one is forcibly reminded of such strange glows from the sky: there are many such listed as having happened on Earth to Earth people. And in view of what we ourselves can demonstrably do to another planet, we have scarcely any right to doubt it could happen to us.

That's one possible explanation. Charles Fort would delight in throwing in a few other possibilities. But to disbelieve anything because for the time being no credible answer can be found is one thing that no one who has read Fort can ever quite do. Because we are beginning to realize now that science is subject to changes internally and externally and that the line between the possible and the impossible is purely arbitrary. It depends on how much your own mind is prepared to bless and how much it too prefers to relegate to the limbo of the forgotten—the damned realms of data ignored by science.

Reading Charles Fort was certainly one of the formative influences of my life, as it has been on the lives of countless tens of thousands of others. From the writings of this acidulous compiler of the *outré* and unexpected, one can only achieve that very necessary widening of the mental boundaries that has become almost vital to stability in this unstable age. It is an education in re-education for the intellectually curious, for the person determined to keep out of the rut of complacency and simple-minded acceptance, for the reader who delights in the thrill of finding out in detail what a really whacky wonderful universe we inhabit.

Certainly the flotillas of flying saucers seen in the past two decades had their home port in the pages of Fort. Certainly a great part of the science-fiction thinking that dominates both the literature of deliberate fantasy and the indeliberate fantasy of Atom Age journalism and politics must owe a good deal, consciously and unconsciously, to Fort's relentless ploughing through pages of newspapers to bolster up his theme that the only sure attitude for a sane mind is to doubt that which is accepted and to accept only that which is subject to doubt.

If this baffles you, it should also serve to intrigue you into reading for yourself *The Book of the Damned*, the work of one of the truly original minds of this century, a book from which you cannot emerge as quite the same person you were when you first started its fascinating pages. It is an experience akin to a needle-shower of the mind—exhilarating, exciting, and endlessly stimulating.

Introduction

by

TIFFANY THAYER

Great books have this in common with the sea and with women—that they are all things to all men. The book in your hand has that in common with all other great books, and to me it is a catalyst. Throw a paragraph of it into any company and stand well back!

This book attracts—as a magnet—the curious-, prying-, inquisitive-minded, and holds those who have affinity or sympathy for its powers. This, too, is seen to be a selective process since not all substances are susceptible to the same attractions and repulsions. Once again—the discrimination is not as to value or to any ultimate whatever but simply as to affinities. Your little red-painted horseshoe magnet ignores gold—or —vice versa. I know not how to tell the actual or absolute specific gravity of ideas. It is no more than my prejudice to think that the weightiest find here their level. Other men: other prejudices.

As a lens turned on life—more particularly on Science and most specifically on Physics and Astronomy—this book, like any other lens, is dependent for any result upon the sensitivities behind it. My own eye—which I have no reason to suppose differed from any when I first applied it to *The Book of the Damned* in 1919—craved the achromatic lens of high power. The film of brain exposed behind it preferred to take expressions as untinted as might be. It reacted to the truth with as little bias as the pack or roll you put in your camera. Other minds: other preferences. Would you see all in a rosy haze—or indigo? Would you? Then drop this book like a hot potato.

Yes to that one group which *knows* what it thinks, knows what it *wishes* to think and *would* not change, this volume is not recommended. One can only regret that group is such a large one.

Before we congratulate each other too cordially on our

esoterism, however, our numbers can be further diminished by eliminating the millions who cannot laugh and think at the same time: that is, all pedants in or out of the classroom who approach either reading or writing as a task rather than a pleasure or who maintain that dignity is an integral part of either teaching or learning—them as well as their mirthless product. For Charles Fort packed a belly laugh in either typewriter hand, as you shall see. He laughed as he wrote, as he read, as he thought: he roared at his subject, guffawed at the pretensions of its serious practitioners, chuckled at his readers, snickered at his correspondents, smiled at his own folly for engaging in such a business, grinned at the reviews of his books and became hilarious at my expense when he saw that I was actually organizing the Fortean Society.

The phrase is so abused as to have lost almost all meaning, but if you can go back to what the words once stood for in their primitive, pristine state—Charles Fort had the most magnificent "sense of humour" that ever made life bearable to a thoughtful man. Never forget that as you read him. If you do, he'll trick you. He'll make you hopping mad sometimes, but—as your choler rises—remember he's doing it purposely and that just when you're boiling he'll stick his head up and thumb his nose at you.... Naturally, such horseplay is frowned upon in professorial circles. I could venture to say why, but if I address adults that is hardly necessary.

Even if it were true that newspapers were Charles Fort's chief source, that judgment is a greater criticism of the daily press than of our author: but it is not true. Research will reveal that so-called "scientific journals" and the publications, transactions, reports, and so on, of learned societies far outnumber the reference to daily papers. His two subsequently published volumes, *Lo!* and *Wild Talents*, both of which are frankly books about people and the strange things that happen to them of which newspapers provide practically the only record extant.... However, let us not quarrel with these unobservant ones but wave them a merry goodbye and Godspeed to the books *they* prefer. Let them beat the drum for Einstein—since only twelve persons in the world are supposed to be capable of understanding him. Perhaps they feel more at home in that company: perhaps they are uncomfortable except in a kneeling position. Mental genuflection is not characteristic of Forteans.

Let us stone from our Temple those who reviled Fort for his effrontery in daring to accuse learned men of charlatanism when he had not a single letter to add after his name. It escapes the attention of these observers that by the time a student attains such a badge of competence in one or another of the Sciences he is almost invariably so thoroughly imbued with the dogmas of that craft, guild or racket that he can no longer see about him with sufficient clarity to think such thoughts as these, much less to write comparable books. It is hardly short of miraculous that we were spared such a mentality as Fort's—gloriously free from all the cant of the schoolmen and with no axe of his own to grind.

That Fort had no axe to grind brings us to another group, they who cannot see the greatness of a Titanic destroyer... *The great thing to remember is that the mind of man cannot be enlightened permanently by merely teaching him to reject some particular set of superstitions*, says Gilbert Murray in his "Four Stages of Greek Religion". *There is an infinite supply of other superstitions always at hand; and the mind that desires such things—that is, the mind that has not trained itself to the hard discipline of reasonableness and honesty, will, as soon as its devils are cast out, proceed to fill itself with their relations....* The plaint of this group is for "something instead". They demand of Charles Fort that he supply them with a faith to supplant the old. If Earth isn't round, what shape is it? If light doesn't get here from the sun in eight minutes, how long *does* it take? If light has no velocity, what does it have? And so on ... Fort gives them as many answers as there are fleas on a dog—breath-taking answers, tongue-in-cheek answers, brilliant answers, colourful, staggering—and all no more, no less, preposterous than the answers they once thought were true.

On the other hand, there are those detractors who classify Fort with the cranks as if he had written to establish some private and well-loved cosmos of his own. They take his celestial argosies seriously. When he says: "I think we're fished for"—they are sufficiently naïve to understand by that that Charles Fort believed *we* are "fished for" ... As an intimate of the man through a period of years, permit me to assure you that he believed nothing of the kind. Yet: prove that we are *not* fished for if you can.

Charles Fort was in no sense a crank. He believed not one hair's breadth of any of his amazing "hypotheses"—as any sensible adult must see from the text itself. He put his theses forward jocularly—as Jehovah must have made the platypus and, perhaps, man. He put them forward for literary purposes—and he achieved them. I once asked him what he called himself—a "neo-astronomer" or a "philosopher" or what? Fort said: "I'm just a writer." And that he was!... At another time I talked about his style, its inimitability, the sheer joy of it, and Fort said: "I wonder if the energy that goes into manner wouldn't be better devoted to matter." And the next day he wrote: *Mineral specimens now in museums—calcites that are piles of petals—or that long ago were the rough notes of a rose.*

One more large group we must eject before the services begin: the group which calls Charles Fort the "arch-enemy of science". Flatly—he was nothing of the kind. The newspapers gave him that label, the reporters and reviewers being unwilling or unable to invent a more accurate term. Nowadays, when living frogs or fishes or anything else which "should" not fall from the sky come down anyway, the papers say: *Here is another datum for that arch-enemy of science, Charles Fort.* This is a quick means of conveying a false impression which has been crystallized by time and repetition: a short-cut for the mentally lazy on both sides of the newspaper column, none of whom—on either side—has any concern for abstract justice or for the honesty of their own thinking.

In this same way and at the hands of these same journalists, Charles Fort is characterized as a marvel-monger, classed with Ripley, *et al.* It is so easy to say: "That fellow who wrote all the books about red snow and black rain." These pithy word-cartoons, unjust and fallacious, are typical of the exchange of misconceptions which pass for information in this world, bringing to mind what the Great Dissenter of the United States Supreme Court, and one of the greatest Forteans, Oliver Wendell Holmes, once said: "Certitude is not the test of certainty: we have been cocksure of many things that were not so."

Charles Fort was the arch-enemy of dogma—not of science—which every line that follows in this volume well

10

attests. If he says, *accept only temporarily*: and that—ideally and theoretically—is the basic principle of the scientific method. As I said before in somewhat different words, it is Fort's greatest glory that he emphasized the paramount necessity for maintaining the *ephemeral* nature of all acceptances until the last dog is hung. Every scientist worth his salt maintains exactly the same thing. Yet Fort is popularly, widely, notoriously known as the "arch-enemy" of this, the first—and perhaps the only—article in his creed.

Ah, well, that's the papers for you. No wonder that one group of Fort's detractors finds it a weakness in him that he quotes from newspapers at all. It was the papers which created Science with a capital "S" that all-highest, all-knowing court of last appeal. The diggers and delvers in the laboratories don't even recognize their own portraits as presented there. But, here I am on dangerous ground. We are glancing at one segment of that group mentioned above—the group which wishes Charles Fort had never written these books. Let us pass on.

Let us sum up:

I call this one of the greatest books ever written in this world, right up there at the top, surely among the first ten. That estimate is based on its potentiality rather than upon any mensurable effect to date. That potentiality lies in its power to *make* its readers think without telling them *what* to think. That is the *avowed* intention of our schools: an intention which, admittedly, they have never achieved. The book is delightfully phrased—something which cannot be said for any text ever handed me by a teacher to study. It encourages the curious to question, the prying to pry, the inquisitive to inquire. Is there a higher mission on earth? I deem not.

When *The Book of the Damned* exploded in the office of the Chicago *Daily News*, Henry Blackman Sell was editor of the Wednesday Book Page, easily the liveliest, healthiest most civilized literary review this country has ever seen, before or since. Carl Sandburg reviewed motion pictures for it. Keith Preston wrote verse, Floyd Dell, Sherwood Anderson and Ben Hecht were contributors. Ben Hecht reviewed Fort's first book. The review is a classic and it was reprinted in full in the *Fortean Society Magazine*,

January, 1940. Here is only a line: "Charles Fort has made a terrible onslaught upon the accumulated lunacy of fifty centuries. The onslaught will perish. The lunacy will survive, intrenching itself behind the derisive laughter of all good citizens."

But the acid of Charles Fort's prose had bitten deep into many of us and we got our heads together to such fashion that his onslaught survived in this world, in spite of the derisive laughter of all good citizens. We founded the Fortean Society—to honour the man who had written: "I conceive of nothing, in religion, science, or philosophy, that is more than the proper thing to wear, for a while."

The Wednesday following publication of Ben Hecht's review of *The Book of the Damned* in the Chicago *Daily News*, this appeared:

WHO IS CHARLES FORT?

After reading "The Book of the Damned" (published by Boni and Liveright) we were keen to know all about the author. A wire brought a wire and here you are, fresh from the Western Union. (New York Special to the Wednesday Book Page.)

I began writing "The Book of the Damned" when I was a boy. I had determined to be a naturalist. I read voraciously, shot birds and stuffed them, collected stamps, classified minerals, stuck insects on pins and labelled them as I saw them labelled in museums. I became a newspaper reporter and instead of collecting idealists' bodies in morgues, Sunday school children parading in Brooklyn, greengoods men and convicts in jail, I arranged my experiences, I pottered over them quite as I had over birds' eggs and minerals and insects.

I am astonished every time I hear one say he cannot understand dreams, or, rather he sees anything especially mystic in dreams. Let alone look back at his own life. There are no dreams phenomena which are not characteristic of all lives—the fading away, the dissolving again of something I had supposed would be final, so interesting, sometimes so exciting were they—fragments of bodies in morgues—crime and altruism. Therein was born the monism that runs through "The Book of the Damned", the merging away of all things in all other things, the impossibility of distinguishing anything from anything else in a

12

positive sense or specifically of distinguishing everyday life from dream existence.

I determined to write a book. I began writing novels. Year in, year out, 3,500,000 words, though that's only an estimate.

I thought that, except in the writing of novels, which probably looked like the offspring of kangaroos, not an incentive could there be to go on living. Lawyers and naturalists and longshoremen and United States Senators—what a dreary lot! But I hadn't written what I wanted. I'd begin anew and be an ultra-scientific realist.

So I took notes enormously. I had a wall covered with pigeon-holes for them. I had 25,000 notes. I worried when I thought of the possibility of fire. I thought of taking the notes upon fireproof material. But they were not what I wanted, and, finally, I destroyed them. For that Theodore Dreiser will never forgive me.

My first interests had been scientific—realism sent me back. Then for eight years I studied all the arts and sciences I had ever heard of and I invented half a dozen arts and sciences. I marvelled that anybody could be satisfied to be a novelist or the head of a steel trust or a tailor or a governor or a street cleaner.

Then came to me a plan of collecting notes upon all subjects of human research upon all known phenomena, and then to try to find the widest possible diversity of data, agreements that would signify something of cosmic order or law or formula, something that could be generalized. I collected notes upon principles and phenomena of astronomy, sociology, psychology, deep sea diving, navigation, surveying, volcanoes, religion, sexes, earthworms—that is, always seeking similarities in widest seeming differences such as astronomic and chemic and sociologic quantivalence or astronomic and chemic sociologic perturbations, chemic and musical combinations, morphologic phenomena of magnetism, chemistry and sexual attractions.

I ended up with 40,000 notes arranged under 1,300 headings such as "Harmony", "Equilibrium", "Catalysts", "Saturation", "Supply and Demand", "Metabolism".

They were 1,300 hell hounds gibing, with 1,300 voices, at my attempt to find finality. I wrote a book that expressed very little of what I was trying to do.

I cut it down from 500 or 600 pages to ninety pages. Then I put it away. It was not what I wanted. But the force of the 40,000 notes had been modified by this book. Nevertheless, the power, or the hypnosis, of them, orthodox notes, all of them, orthodox materialism, Tyndall says this, Darwin says that, authoritativeness, positiveness, chemists and astronomers and geologists have proved this or that, nevertheless, monism and revolt were making me write that not even are twice two four, except arbitrarily and conventionally, or that there is no positiveness—that the most profoundly hypnotized subject has some faint awareness of his state—and that, with a doubt here and a dissatisfaction there, I never had been any more absolutely faithful to scientific orthodoxy than ever has been mediaeval monk or member of the Salvation Army—altogether unquestioning. The oneness of allness. That in my attempt to find the underlying in all phenomena I had been wrong in the two classifications that I had ended up with—that these two orders of appearance represent ideal extremes that have no existence in our state of seeming, that we and all other appearances of phantasms in a superdream are expressions of one cosmic flow or graduation between them; one called disorder, unreality, inequilibrium, ugliness, discord, inconsistence; the other called order, realness, equilibrium, beauty, harmony, justice, truth. This is the underlying theme in "The Book of the Damned". It is something that many persons have not wanted.

CHARLES FORT

Charles Hoy Fort was born August 9, 1874, in Albany, New York. He died May 3, 1932, in Royal Hospital, The Bronx, New York. He was nearly if not quite six feet tall, heavy, fair. He wore a brown moustache that bristled somewhat less than Nietzsche's. His sight was failing in his last years and his glasses had to be thick-lensed. He was an anachronism in modern dress incongruous in his Bronx flat. As we sat with some home brew of his making, strong cheeses, coarse rye bread and "whiskied grapes" at a circular dining-table, talking the night away, it often occurred to me that his frame called for leather and buckles, that the board should have been bare and brown, washed by slops from heavy tankards and worn smooth by heavy swordhands. The light should have been from flambeaux and—to match our words—Faust and Villon should have stopped by in passing on their way to murder or conference with the devil.

Lacking all those props we made our own atmosphere with high talk and set new worlds to spinning in it, finding godhead a simple business, quite easy of achievement and within the reach of all.

Fort had neither friends nor acquaintances. The only other person he welcomed there was Theodore Dreiser. Mrs. Fort—Anna—attended the neighbourhood movies every night, often with "Charlie". (How he hated to be called "Charlie"!) She lamented his unsocial bent, knew all her neighbours' affairs, attended the family food and drink with skill and some imagination and never dreamed what went on in her husband's head, never read his or any other books. She survived him a little more than five years.

On the walls of the flat were framed specimens of giant spiders, butterflies, weird creatures adept at concealment, imitating the sticks and leaves to which they were affixed. There was also framed a photograph of a baseball beside a hailstone, both objects the same size, sent to Fort by a correspondent, and—under glass—a specimen of some stuff that looks like dirty, shredded asbestos which had fallen from the sky in quantities covering several acres.

In all other respects the domicile was quite commonplace, the sort of home indicated on theatrical scene-plots by the phrase "shabby-genteel". Fort was—to the extent of his simple needs—independent of any income from his writing. In that Bronx flat he wrote *Lo!* and *Wild Talents*, the latter published only a week or two after his death. There he invented Super-checkers, a game played with *armies* of men on a vast board with hundreds of squares. In a letter to me he said: "Super-checkers is going to be a great success. I have met four more persons who consider it preposterous."

What more shall I say of Charles Fort personally? He was one sane man in a mad, mad world—and for that reason very lonely. He had not a single illusion, not even about himself or his work. Anything more I might say is implicit in the pages which follow. He put himself on paper—gorgeously, uproariously, with gusto.

Chapter 1

A procession of the damned.

By the damned, I mean the excluded.

We shall have a procession of data that Science has excluded.

Battalions of the accursed, captained by pallid data that I have exhumed, will march. You'll read them—or they'll march. Some of them livid and some of them fiery and some of them rotten.

Some of them are corpses, skeletons, mummies, twitching, tottering, animated by companions that have been damned alive. There are giants that will walk by, though sound asleep. There are things that are theorems and things that are rags: they'll go by like Euclid arm in arm with the spirit of anarchy. Here and there will flit little harlots. Many are clowns. But many are of the highest respectability. Some are assassins. There are pale stenches and gaunt superstitions and mere shadows and lively malices: whims and amiabilities. The naïve and the pedantic and the bizarre and the grotesque and the sincere and the insincere, the profound and the puerile.

A stab and a laugh and the patiently folded hands of hopeless propriety.

The ultra-respectable, but the condemned, anyway.

The aggregate appearance is of dignity and dissoluteness: the aggregate voice is a defiant prayer: but the spirit of the whole is professional.

The power that has said to all things that they are damned, is Dogmatic Science.

But they'll march.

The little harlots will caper, and freaks will distract attention, and the clowns will break the rhythm of the whole with their buffooneries—but the solidity of the procession as a whole: the impressiveness of things that pass and pass and pass, and keep on and keep on and keep on coming.

The irresistibleness of things that neither threaten nor jeer nor defy, but arrange themselves in mass-formations that pass and pass and keep on passing.

So, by the damned, I mean the excluded.

By the excluded I mean that which will some day be the excluding.

Or everything that is, won't be.

And everything that isn't, will be—

But of course, will be that which won't be—

It is our expression that the flux between that which isn't and that which won't be, or the state that is commonly and absurdly called "existence", is a rhythm of heavens and hells: that the damned won't stay damned; that salvation only precedes perdition. The inference is that some day our accursed tatterdemalions will be sleek angels. Then the subinference is that some later day, back they'll go back whence they came.

It is our expression that nothing can attempt to be, except by attempting to exclude something else: that that which is commonly called "being" is a state that is wrought more or less definitely proportionately to the appearance of positive difference between that which is included and that which is excluded.

But it is our expression that there are no positive differences: that all things are like a mouse and a bug in the heart of a cheese. Mouse and a bug: no two things could seem more unlike. They're there a week, or they stay there a month: both are then only transmutations of cheese. I think we're all bugs and mice, and are only different expressions of an all-inclusive cheese.

Or that red is not positively different from yellow: is only another degree of whatever vibrancy yellow is a degree of: that red and yellow are continuous, or that they merge in orange.

So then that, if, upon the basis of yellowness and redness, Science should attempt to classify all phenomena, including all red things as veritable, and excluding all yellow things as false or illusory, the demarcation would have to be false and arbitrary, because things coloured orange, constituting continuity, would belong on both sides of the attempted border-line.

As we go along, we shall be impressed with this:

That no basis for classification, or inclusion and exclusion, more reasonable than that of redness and yellowness has ever been conceived of.

Science has, by appeal to various bases, included a mul-

18

titude of data. Had it not done so, there would be nothing with which to seem to be. Science has, by appeal to various bases, excluded a multitude of data. Then, if redness is continuous with yellowness: if every basis of admission is continuous with every basis of exclusion, Science must have excluded some things that are continuous with the accepted. In redness and yellowness, which merge in orangeness, we typify all tests, all standards, all means of forming an opinion—

Or that any positive opinion upon any subject is illusion built upon the fallacy that there are positive differences to judge by—

That the quest of all intellection has been for something—a fact, a basis, a generalization, law, formula, a major premise that is positive: that the best that has ever been done has been to say that some things are self-evident—whereas, by evidence we mean the support of something else—

That this is the quest; but that it has never been attained; but that Science has acted, ruled, pronounced, and condemned as if it had been attained.

What is a house?

It is not possible to say what anything is, as positively distinguished from anything else, if there are no positive differences.

A barn is a house, if one lives in it. If residence constitutes houseness, because style of architecture does not, then a bird's nest is a house: and human occupancy is not the standard to judge by, because we speak of dogs' houses; not material, because we speak of snow houses of Eskimos—or a shell is a house to a hermit crab—or was to the mollusc that made it—or things seemingly so positively different as the White House at Washington and a shell on the seashore are seen to be continuous.

So no one has ever been able to say what electricity is, for instance. It isn't anything, as positively distinguished from heat or magnetism or life. Metaphysicians and the theologians and biologists have tried to define life. They have failed, because, in a positive sense, there is nothing to define: there is no phenomenon of life that is not, to some degree, manifest in chemism, magnetism, astronomic motions.

White coral islands in a dark blue sea.

Their seeming of distinctness: the seeming of individuality, or of positive difference one from another—but all are only projections from the same sea bottom. The difference between sea and land is positive. In all water there is some earth: in all earth there is some water.

So then that all seeming things are not things at all, if all are inter-continuous, any more than is the leg of a table a thing in itself. It is only a projection from something else: that not one of us is a real person, if, physically, we're continuous with environment; if, psychically, there is nothing to us but expression of relation to environment.

Our general expression has two aspects:

Conventional monism, or that "all things" that seem to have identity of their own are only islands that are projections from something underlying, and have no real outlines of their own.

But that all "things", though only projections, are projections that are striving to break away from the underlying that denies them identity of their own.

I conceive of one inter-continuous nexus, in which and of which all seeming things are only different expressions, but in which all things are localizations or one attempt to break away and become real things, or to establish entity of positive difference or final demarcation or unmodified independence—or personality or soul, as it is called in human phenomena—

That anything that tries to establish itself as a real or positive, or absolute system, government, organization, self, soul, entity, individuality, can so attempt only by drawing a line about itself, or about the inclusions that constitute itself, and damning or excluding, or breaking away from, all other "things":

That, if it does not so act, it cannot seem to be;

That, if it does so act, it falsely and arbitrarily and futilely and disastrously acts, just as would one who draws a circle in the sea, including a few waves, saying that the other waves, with which the included are continuous, are positively different, and stakes his life upon maintaining that the admitted and the damned are positively different.

Our expression is that our whole existence is animation of the local by an ideal that is realizable only in the universal:

That, if all exclusions are false, because always are included and excluded continuous: that if all seeming of existence perceptible to us is the product of exclusion, there

20

is nothing that is perceptible to us that really is: that only the universal can really be.

Our especial interest is in modern science as a manifestation of this one ideal or purpose or process:

That it has falsely excluded, because there are no positive standards to judge by: that it has excluded things that, by its own pseudo-standards, have as much right to come in as have the chosen.

Our general expression:

That the state that is commonly and absurdly called "existence", is a flow, or a current, or an attempt, from negativeness to positiveness, and is intermediate to both.

By positiveness we mean:

Harmony, equilibrium, order, regularity, stability, consistency, unity, realness, system, government, organization, liberty, independence, soul, self, personality, entity, individuality, truth, beauty, justice, perfection, definiteness—

That all that is called development progress, or evolution is movement toward, or attempt toward, this state for which, or for aspects of which, there are so many names, all of which are summed up in the one word "positiveness".

At first this summing up may not be very readily acceptable. At first it may seem that all these words are not synonyms: that "harmony" may mean "order", but that by independence, for instance, we do not mean "truth", or that by "stability" we do not mean "beauty", or "system", or "justice".

I conceive of one inter-continuous nexus, which expresses itself in astronomic phenomena, and chemic, biologic, psychic, sociologic: that it is everywhere striving to localize positiveness: that to this attempt in various fields of phenomena—which are only quasi-different—we give different names. We speak of the "system" of the planets and not of their "government": but in considering a store for instance, and its management, we see that the words are interchangeable. It used to be customary to speak of chemic equilibrium, but not of social equilibrium: that false demarcation has been broken down. We shall see that by all these words we mean the same state, as every-day conveniences, or in terms of common illusions, of course, they are not synonyms. To a child an earth worm is not an animal. It is to the biologist.

By "beauty", I mean that which seems complete.

21

Obversely, that the incomplete, or the mutilated, is the ugly. Venus de Milo.

To a child she is ugly.

When a mind adjusts to thinking of her as a completeness, even though, by physiologic standards, incomplete, she is beautiful.

A hand thought of only as a hand, may seem beautiful. Found on a battlefield—obviously a part—not beautiful.

But everything in our experience is only a part of something else that in turn is only a part of still something else—or that there is nothing beautiful in our experience: only appearances that are intermediate to beauty and ugliness—that only universality is complete: that only the complete is beautiful: that every attempt to give to the local the attribute of the universal.

By stability, we mean the immovable and the unaffected. But all seeming things are only reactions to something else. Stability, too then, can be only the universal, or that besides which there is nothing else. Though some things seem to have—or have—higher approximations to stability than have others, there are, in our experience, only various degrees of intermediateness to stability and instability. Every man, then, who works for stability under its various names of "permanency", "survival", "duration", is striving to localize in something the state that is realizable only in the universal.

By independence, entity, and individuality, I can mean only that besides which there is nothing else, if given only two things, they must be continuous and mutually affective, if everything is only a reaction to something else, and any two things would be destructive of each other's independence, entity, or individuality.

All attempted organizations and systems and consistencies, some approximating far higher than others, but all only intermediate to Order and Disorder, fail eventually because of their relations with outside forces. All are attempted completenesses. If to all local phenomena there are always outside forces, these attempts, too, are realizable only in the state of completeness, or that to which there are no outside forces.

Or that all these words are synonyms, all meaning the state that we call the positive state—

That our whole "existence" is a striving for the positive

state.

The amazing paradox of it all:

That there is only this one process, and that it does animate all expressions, in all fields of phenomena, of that which we think of as one inter-continuous nexus:

The religious and their idea or deal of the soul. They mean distinct, stable entity, or a state that is independent, and not a mere flux of vibrations or complex of reactions to environment, continuous with environment, merging away with an infinitude of other independent complexes.

But the only thing that would not merge away into something else would be that besides which there is nothing else.

That Truth is only another name for the positive state, or that the quest for Truth is the attempt to achieve positiveness:

Scientists who have thought that they were seeking Truth, but who were trying to find out astronomic, or chemic, or biologic truths. But Truth is that besides which there is nothing: nothing to modify it, nothing to question it, nothing to form an exception: the all-inclusive, the complete—

By Truth I mean the Universal.

So chemists have sought the true, or the real, and have always failed in their endeavours, because of the outside relations of chemical phenomena: have failed in the sense that never has a chemical law, without exceptions, been discovered: because chemistry is continuous with astronomy, physics, biology—For instance, if the sun should greatly change its distance from this earth, and if human life could survive, the familiar chemic formulas would no longer work out: a new science of chemistry would have to be learned—

Or that all attempts to find Truth in the special are attempts to find the universal in the local.

And artists and their striving for positiveness, under the name of "harmony"—but their pigments that are oxydizing, or are responding to a deranging environment—or the strings of musical instruments that are differently and disturbingly adjusting to outside chemic and thermal and gravitational forces—again and again this oneness of all ideals, and that it is the attempt to be, or to achieve locally, that which is realizable only universally. In our experience there is only intermediateness to harmony and discord. Harmony is that besides which there are no outside forces.

And nations that have fought with only one motive: for

23

individuality, or entity, or to be real, final nations, not subordinate to, or parts of, other nations. And that nothing but intermediateness has ever been attained, and that history is record of failures of this attempt, because there always have been outside forces, or other nations contending for the same goal.

As to physical things, chemic, mineralogic, astronomic, it is not customary to say that they act to achieve Truth or Entity, but it is understood that all motions are toward Equilibrium: that there is no motion except toward Equilibrium, of course always away from some other approximation to Equilibrium.

All biologic phenomena act to adjust: there are no biologic actions other than adjustments.

Adjustment is another name for Equilibrium. Equilibrium is the Universal, or that which has nothing external to derange it.

But that all that we call "being" is motion: and that all motion is the expression, not of equilibrium, but of equilibrating, or of equilibrium unattained: that life-motions are expressions of equilibrium unattained: that all thought relates to the unattained: that to have what is called being in our quasi-state, is not to be in the positive sense, or is to be intermediate to Equilibrium and Inequilibrium.

So then:

That all phenomena in our intermediate state, or quasi-state, represent this one attempt to organize, stabilize, harmonize, individualize—or to positivize, or to become real:

That only to have seeming is to express failure or intermediateness to final failure and final success:

That every attempt—that is observable—is defeated by Continuity, or by outside forces—or by the excluded that are continuous with the included:

That our whole "existence" is an attempt by the relative to be the absolute, or by the local to be the universal.

In this book, my interest is in this attempt as manifest in modern science:

That it has attempted to be real, true, final, complete, absolute:

That, if the seeming of being, here, in our quasi-state, is the product of exclusion that is always false and arbitrary, if always are included and excluded continuous, the whole seeming system, or entity, of modern science is only quasi-

24

system, or quasi-entity, wrought by the same false and arbitrary process as that by which the still less positive system that preceded it, or the theological system, wrought the illusion of its being.

In this book, I assemble some of the data that I think are of the falsely and arbitrarily excluded.

The data of the damned.

I have gone into the outer darkness of scientific and philosophical transactions and proceedings, ultra-respectable but covered with the dust of disregard. I have descended into journalism. I have come back with the quasi-souls of lost data.

They will march.

As to the logic of our expressions to come—

That there is only quasi-logic in our mode of seeming:

That nothing ever has been proved—

Because there is nothing to prove.

When I say that there is nothing to prove, I mean that to those who accept Continuity, or the merging away of all phenomena into other phenomena, without positive demarcations one from another, there is, in a positive sense, no one thing. There is nothing to prove.

For instance nothing can be proved to be an animal—because animalness and vegetableness are not positively different. There are expressions of life that are as much vegetable as animal, or that represent the merging of animalness and vegetableness. There is then no positive test, standard, criterion, means of forming an opinion. As distinct from vegetables, animals do not exist. There is nothing to prove. Nothing could be proved to be good, for instance. There is nothing in our "existence" that is good, in a positive sense, or as really outlined from evil. If to forgive be good in time of peace, it is evil in wartime. There is nothing to prove: good in our experience is continuous with, or is only another aspect of evil.

As to what I'm trying to do now—I accept only. If I can't see universally, I only localize.

So, of course then, that nothing ever has been proved:

That theological pronouncements are as much open to doubt as ever they were, but that, by a hypnotizing process, they became dominant over the majority of minds in their era;

That, in a succeeding era, the laws, dogmas, formulas,

25

principles, of materialistic science never were proved, because they are only localizations simulating the universal; but that the leading minds of their era of dominance were hypnotized into more or less firmly believing them.

Newton's three laws, and that they are attempts to achieve positiveness, or to defy and break Continuity, and are as unreal as are all other attempts to localize the universal:

That, if every observable body is continuous, mediately or immediately, with all other bodies, it cannot be influenced only by its own inertia, so that there is no way of knowing what the phenomena of inertia may be; that, if all things are reacting to an infinitude of forces there is no way of knowing what the effects of only one impressed force would be; that if every reaction is continuous with its action, it cannot be conceived of as a whole, and that there is no way of conceiving what it might be equal and opposite to—

Or that Newton's three laws are three articles of faith;

Or that demons and angels and inertias and reactions are all mythological characters;

But that, in their eras of dominance, they were almost as firmly believed in as if they had been proved.

Enormities and preposterousnesses will march.

They will be "proved" as well as Moses or Darwin or Lyell ever "proved" anything.

We substitute acceptance for belief.

Cells of an embryo take on different appearances in different eras.

That social organism is embryonic.

The more firmly established, the more difficult to change.

That firmly to believe is to impede development.

That only temporarily to accept is to facilitate.

But:

Except that we substitute acceptance for belief, our methods will be the conventional methods; the means by which every belief has been formulated and supported: or our methods will be the methods of theologians and savages and scientists and children. Because, if all phenomena are continuous, there can be no positively different methods. By the inconclusive means and methods of cardinals and fortune tellers and evolutionists and peasants, methods

26

which must be inconclusive, if they relate always to the local, and if there is nothing local to conclude, we shall write this book.

If it function as an expression of its era it will prevail.

All sciences begin with attempts to define.

Nothing ever has been defined.

Because there is nothing to define.

Darwin wrote *The Origin of Species*.

He was never able to tell what he meant by a "species".

It is not possible to define.

Nothing has ever been finally found out.

Because there is nothing final to find out.

It's like looking for a needle that no one ever lost in a haystack that never was—

But that all scientific attempts really to find out something, whereas really there is nothing to find out, are attempts, themselves, really to be something.

A seeker of Truth. He will never find it. But the dimmest of possibilities—he may himself become Truth.

Or that science is more than an inquiry:

That it is a pseudo-construction, or a quasi-organization: that it is an attempt to break away and locally establish harmony, stability, equilibrium, consistency, entity—

Dimmest of possibilities—that it may succeed.

That ours is a pseudo-existence and all appearances in it partake of its essential fictitiousness—

But that some appearances approximate far more highly to the positive state than do others.

We conceive of all "things" as occupying gradations, or steps in series between positiveness and negativeness or realness and unrealness: that some seeming things are more nearly consistent, just, beautiful, unified, individual, harmonious, stable—than others.

We are not realists. We are not idealists. We are intermediatists—that nothing is real but that nothing is unreal: that all phenomena are approximations one way or the other between realness and unrealness.

So then:

That our whole quasi-existence is an intermediate stage between positiveness and negativeness or realness and unrealness.

27

Like purgatory, I think.

But in our summing up, which was very sketchily done, we omitted to make clear that Realness is an aspect of the positive state.

By Realness, I mean, that which does not merge away into something else and that which is not partly something else: that which is not a reaction to, or an imitation of, something else. By a real hero, we mean one who is not partly a coward, or whose action and motives do not merge away into cowardice. But, if in Continuity, all things do merge, by Realness, I mean the Universal, besides which there is nothing with which to merge.

That, though the local might be universalized, it is not conceivable that the universal can be localized: but that high approximations there may be, and that these approximate successes may be translated out of Intermediateness into Realness—quite as, in a relative sense, the industrial world recruits itself by translating out of unrealness, or out of the seemingly less real imaginings of inventors, machines which seem, when set up in factories, to have more of Realness than they had when only imagined.

That all progress, if all progress is toward stability, organization, harmony, consistency, or positiveness, is the attempt to become real.

So, then, in general metaphysical terms, our expression is that like a purgatory, all that is commonly called "existence", which we call Intermediateness is quasi-existence neither real nor unreal, but expression of attempt to become real, or to generate for or recruit a real existence.

Our acceptance is that Science, though usually thought of so specifically, or in its own terms, usually supposed to be a prying into old bones, bugs, unsavoury messes, is an expression of this one spirit animating all Intermediateness: that, if Science could absolutely exclude all data but its own present data, or that which is assimilable with the present quasi-organization, it would be a real system, with positively definite outlines—it would be real.

Its seeming approximation to consistency, stability, system—positiveness or realness—is sustained by damning the irreconcilable of the unassimilable—

All would be well.

All would be heavenly—

If the damned would only stay damned.

28

Chapter 2

In the autumn of 1883, and for years afterward, occurred brilliant-coloured sunsets, such as had never been seen before within the memory of all observers. Also there were blue moons.

I think that one is likely to smile incredulously at the notion of blue moons. Nevertheless they were common as were green suns in 1883.

Science had to account for these unconventionalities. Such publications as *Nature* and *Knowledge* were besieged with inquiries.

I suppose, in Alaska and in the South Sea Islands, all the medicine men were similarly upon trial.

Something had to be thought of.

Upon the 28th of August, 1883, the volcano of Krakatoa, of the Straits of Sunda, had blown up.

Terrific.

We're told that the sound was heard 2,000 miles, and that 36,380 persons were killed. Seems just a little unscientific or impositive, to me: marvel to me we're not told 2,163 miles and 36,387 persons. The volume of smoke that went up must have been visible to other planets—or, tormented with our crawlings and scurryings, the earth complained to Mars; swore a vast black oath at us.

It is said that these phenomena were caused by particles of volcanic dust that were cast high in the air by Krakatoa.

In all text-books that mention this occurrence—no exception so far so I have read—it is said that the extraordinary atmospheric effects of 1883 were first noticed in the last of August or the first of September.

That makes a difficulty for us.

This is the explanation that was agreed upon in 1883—

But for seven years the atmospheric phenomena continued—

Except that, in the seven, there was a lapse of several years—and where was the volcanic dust all that time?

You'd think that such a question as that would make trouble?

Then you haven't studied hypnosis. You have never tried to demonstrate to a hypnotic that a table is not a hippopotamus. According to our general acceptance, it would be impossible to demonstrate such a thing. Point out a hundred reasons for saying that a hippopotamus is not a table: you'll have to end up agreeing that neither is a table—it only seems to be a table. Well, that's what the hippopotamus seems to be. So how can you prove that something is not something else, when neither is something else some other thing? There's nothing to prove.

This is one of the profundities that we advertised in advance.

You can oppose an absurdity only with some other absurdity. But Science is established preposterousness. We divide all intellection: the obviously preposterousness and the established.

But Krakatoa: that's the explanation that the scientists gave. I don't know what whopper the medicine men told.

We see, from the start, the very strong inclination of science to deny, as much as it can, external relations of this earth.

This book is an assemblage of data of external relations of this earth. We take the position that our data have been damned, upon no consideration for individual merits or demerits, but in conformity with a general attempt to hold out for isolation of this earth. This is attempted positiveness. We take the position that science can no more succeed than, in a similar endeavour, could the Chinese, or than could the United States. So then, with only pseudo-consideration of the phenomena of 1883, or as an expression of positivism in its aspect of isolation, or unrelatedness, scientists have perpetrated such an enormity as suspension of volcanic dust seven years in the air—disregarding the lapse of several years—rather than to admit the arrival of dust from somewhere beyond this earth. Not that scientists themselves have ever achieved positiveness, in its aspect of unitedness, among themselves—because Nordenskiold, before 1883, wrote a great deal upon his theory of cosmic dust, and Prof. Cleaveland Abbe contended against the Krakatoan explanation—but that this is the orthodoxy of the main body of scientists.

My own chief reason for indignation here:

That this preposterous explanation interferes with some of my own enormities.

It would cost me too much explaining, if I should have to admit that this earth's atmosphere has such sustaining power.

Later, we shall have data of things that have gone up in the air and that have stayed up—somewhere—weeks—months—but not by the sustaining power of this earth's atmosphere. For instance, the turtle of Vicksburg. It seems to me that it would be ridiculous to think of a good-sized turtle hanging, for three or four months, upheld only by the air, over the town of Vicksburg. When it comes to the horse and the barn—I think that they'll be classics some day, but I can never accept that a horse and a barn could float several months in this earth's atmosphere.

The orthodox explanation:

See the *Report of the Krakatoa Committee of the Royal Society*. It comes out absolutely for the orthodox explanation—absolutely and beautifully, also expensively. There are 492 pages in the "Report", and 40 plates, some of them marvellously coloured. It was issued after an investigation that took five years. You couldn't think of anything done more efficiently, artistically, authoritatively. The mathematical parts are especially impressive: distribution of the dust of Krakatoa; velocity of translation and rates of subsistence; altitudes and persistences—

Annual Register, 1883–105:

That the atmospheric effects that have been attributed to Krakatoa were seen in Trinidad before the eruption occurred;

Knowledge, 5–418:

That they were seen in Natal, South Africa, six months before the eruption.

Inertia and its inhospitality.

Or raw meat should not be fed to babies.

We shall have a few data initiatorily.

I fear me that the horse and the barn were a little extreme for our budding liberalities.

The outrageous is the reasonable, if introduced politely.

Hailstones, for instance. One reads in the newspapers of hailstones the size of hens' eggs. One smiles. Nevertheless I will engage to list one hundred instances, from *Monthly Weather Review*, of hailstones the size of hens' eggs. There is an account in *Nature*, Nov. 1, 1894, of hailstones that weighed almost two pounds each. See Chambers' Encyclopedia for three-pounders. *Report of the Smithsonian Institution*, 1870–479 two-pounders authenticated, and six-pounders reported. At Seringapatam, India, about the year 1800, fell a hailstone—

I fear me, I fear me: this is one of the profoundly damned. I blurt out something that should, perhaps, be withheld for several hundred pages—but that damned thing was the size of an elephant.

We laugh.

Or snowflakes. Size of saucers. Said to have fallen at Nashville, Tenn., Jan. 24, 1891. One smiles.

"In Montana, in the winter of 1887, fell snowflakes 15 inches across, and 8 inches thick," (*Monthly Weather Review*, 1916–73).

In the topography of intellection, I should say that what we call knowledge is ignorance surrounded by laughter.

Black rains—red rains—the fall of a thousand tons of butter.

Jet-black snow—pink snow—blue hailstones—hailstones flavoured like oranges.

Punk and silk and charcoal.

About one hundred years ago, if anyone was so credulous as to think that stones had ever fallen from the sky, he was reasoned with:

In the first place there are no stones in the sky:

Therefore no stones can fall from the sky.

Or nothing more reasonable or scientific or logical than that could be said upon any subject. The only trouble is the universal trouble: that the major premise is not real, or is intermediate somewhere between realness and unrealness.

In 1772, a committee, of whom Lavoisier was a member, was appointed by the French Academy, to investigate a report that a stone had fallen from the sky at Luce, France. Of all attempts at positiveness, in its aspect of isolation, I don't know of anything that has been fought harder for than the notion of this earth's unrelatedness. Lavoisier analysed the stone of Luce. The exclusionists' explanation at that time was that stones do not fall from the sky: that luminous objects may seem to fall, and that hot stones may be picked up where a luminous object seemingly had landed—only lightning striking a stone, heating, even melting it.

The stone of Luce showed signs of fusion.

Lavoisier's analysis "absolutely proved" that this stone had not fallen: that it had been struck by lightning.

So, authoritatively, falling stones were damned. The stock means of exclusion remained the explanation of lightning that was seen to strike something—that had been upon the ground in the first place.

But positiveness and the fate of every positive statement. It is not customary to think of damned stones raising an outcry against a sentence of exclusion, but, subjectively, areolites did—or data of them bombarded the walls raised against them—

Monthly Review, 1796–426

"The phenomenon which is the subject of the remarks before us will seem to most persons as little of worthy credit as any that could be offered. The falling of large stones from the sky, without any assignable cause of their previous ascent, seems to partake so much of the marvellous as almost entirely to exclude the operation of known and natural agents. Yet a body of evidence is here brought to prove that such events have actually taken place, and we ought not to withhold from it a proper degree of attention."

The writer abandons the first, or absolute, exclusion, and modifies it with the explanation that the day before a reported fall of stones in Tuscany, June 16, 1794, there had been an eruption of Vesuvius—

Or that stones do fall from the sky, but they are stones that have been raised to the sky from some other part of the earth's surface by whirlwinds or by volcanic action.

It's more than one hundred and twenty years later. I know of no aerolite that has ever been acceptably traced to terrestrial origin.

Falling stones had to be undamned—though still with a reservation that held out for exclusion of outside forces.

One may have the knowledge of a Lavoisier, and still not be able to analyse, not be able even to see, except conformably with the hypnoses, or the conventional reactions against hypnoses, of one's era.

We believe no more.

We accept.

Little by little the whirlwind and volcano explanations

had to be abandoned, but so powerful was this exclusion-hypnosis, sentence of damnation, or this attempt at positiveness, that far into our own times some scientists, notably Prof. Lawrence Smith and Sir Robert Ball, continued to hold out against all external origins, asserting that nothing could fall to this earth, unless it had been cast up or whirled up from some other part of this earth's surface.

It's as commendable as anything ever has been—by which I mean it's intermediate to the commendable and the censurable.

It's virginal.

Meteorites, data of which were once of the damned, have been admitted, but the common impression of them is only a retreat of attempted exclusion: that only two kinds of substance fall from the sky: metallic and stony: that the metallic are of iron and nickel—

Butter and paper and wool and silk and resin.

We see, to start with, that the virgins of science have fought and wept and screamed against external relations—upon two grounds:

There in the first place;

Or up from one part of this earth's surface and down to another.

As late as November, 1902, in *Nature Notes*, 13–231, a member of the Selborne Society still argued that meteorites do not fall from the sky; that they are masses of iron upon the ground "in the first place", that attract lightning; that the lightning is seen, and is mistaken for a falling, luminous object—

By progress we mean rape.

Butter and beef and blood and a stone with strange inscriptions upon it.

Chapter 3

So then, it is our expression that Science relates to real knowledge no more than does the growth of a plant, or the organization of a department store, or the development of a nation: that all are assimilative, or organizing, or systematizing processes that represent different attempts to attain the positive state—the state commonly called heaven, I suppose I mean.

There can be no real science where there are indeterminate variables, but every variable is, in finer terms, indeterminate, or irregular, if only to have the appearance of being in Intermediateness is to express regularity unattained. The invariable, or the real and stable, would be nothing at all in Intermediateness—rather as, but in relative terms, an undistorted interpretation of external sounds in the mind of a dreamer could not continue to exist in a dreaming mind, because that touch of relative realness would be of awakening and not of dreaming. Science is the attempt to awaken to realness, wherein it is attempt to find regularity and uniformity. Or the regular and uniform would be that which has nothing external to disturb it. By the universal we mean the real. Or the notion is that the underlying super-attempt, as expressed in Science, is indifferent to the subject-matter of Science: that the attempt to regularize is the vital spirit. Bugs and stars and chemical messes: that they are only quasi-real, and that of them there is nothing real to know; but that systematization of pseudo-data is approximation to realness or final awakening—

Or a dreaming mind—and its centaurs and canary birds that turn into giraffes—there could be no real biology upon such subjects, but attempt, in a dreaming mind to systematize such appearances would be movement toward awakening—if better mental co-ordination is all that we mean by the state of being awake—relatively awake.

So it is, that having attempted to systematize, by ignoring externality, to the greatest possible degree, the notion of things dropping in upon this earth, from externality, is as unsettling and as unwelcome to Science as—tin horns blowing in upon a musician's relatively symmetric composition—flies alighting upon a painter's attempted harmony, and tracking colours one into another—suffragist getting up and making a political speech at a prayer meeting.

If all things are of a oneness, which is a state intermediate to unrealness and realness, and if nothing has succeeded in breaking away and establishing entity for itself, and could not continue to "exist" in intermediateness, if it should succeed, any more than could be born still at the same time be the uterine, I of course know of no positive difference between Science and Christian Science—and the attitude of both toward the unwelcome is the same—"it does not exist".

A Lord Kelvin and a Mrs. Eddy, and something not to their liking—it does not exist.

Of course not, we Intermediates say: but, also, that, in Intermediateness, neither is there absolute non-existence.

Or a Christian Scientist and a toothache—neither exists in the final sense: also neither is absolutely non-existent, and according to our therapeutics, the one that more highly approximates to realness will win.

A secret of power—

I think it's another profundity.

Do you want power over something?

Be more nearly real than it.

We'll begin with yellow substances that have fallen upon this earth: we'll see whether our data of them have a higher approximation to realness than have the dogmas of those who deny their existence—that is, as products from somewhere external to this earth.

In mere impressionism we take our stand. We have no positive tests nor standards. Realism in art: realism in science—they pass away. In 1859, the thing to do was to accept Darwinism; now many biologists are revolting and trying to conceive of something else. The thing to do was to accept it in its day, but Darwinism of course was never proved:

The fittest survive.

What is meant by the fittest?

Not the strongest; not the cleverest—

Weakness and stupidity everywhere survive.

There is no way of determining fitness except in that a

thing does survive.

"Fitness," then, is only another name for "survival".

Darwinism:

That survivors survive.

Although Darwinism, then, seems positively baseless, or absolutely irrational, its massing of supposed data, and its attempted coherence approximate more highly to Organization and Consistency than did the inchoate speculations that preceded it.

Or that Columbus never proved that the earth is round.

Shadow of the earth on the moon?

No one has ever seen it in its entirety. The earth's shadow is much larger than the moon. If the periphery of the shadow is curved—but the convex moon—a straight-edged object will cast a curved shadow upon the surface that is convex.

All the other so-called proofs may be taken up in the same way. It was impossible for Columbus to prove that the earth is round. It was not required: only that with a higher seeming of positiveness than that of his opponents, he should attempt. The thing to do, in 1492, was nevertheless to accept that beyond Europe, to the west, were other lands.

I offer for acceptance, as something concordant with the spirit of this first quarter of the twentieth century, the expression that beyond this earth are—other lands—from which come things as, from America, float things to Europe.

As to yellow substances that have fallen upon this earth, the endeavour to exclude extra-mundane origins is the dogma that all yellow rains and yellow snows are coloured with pollen from this earth's pine trees. *Symons' Meteorological Magazine* is especially prudish in this respect and regards as highly improper all advances made by other explainers.

Nevertheless, the *Monthly Weather Review*, May 1887, reports a golden-yellow fall, of Feb. 27, 1877, at Peckloh, Germany, in which four kinds of organisms not pollen, were the colouring matter. There were minute things shaped like arrows, coffee beans, horns, and discs.

They may have been symbols. They may have been objective hieroglyphics—

Mere passing fancy—let it go—

37

In the *Annales de Chimie*, 85–288, there is a list of rains said to have contained sulphur. I have thirty or forty other notes. I'll not use one of them. I'll admit that every one of them is upon a fall of pollen. I said, to begin with, that our methods would be the methods of theologians and scientists, and they always begin with an appearance of liberality. I grant thirty or forty points to start with. I'm as liberal as any of them—or that my liberality won't cost me any-thing—the enormousness of the data that we shall have.

Or just to look over a typical instance of this dogma, and the way it works out:

In the *American Journal of Science*, 1–42–196, we are told of a yellow substance that fell by the bucketful upon a vessel, one "windless" night in June, in Pictou Harbour, Nova Scotia. The writer analysed the substance, and it was found to "give off nitrogen and ammonia and an animal odour".

Now, one of our Intermediatist principles, to start with, is that so far from positive, in the aspect of Homogeneous-ness, are all substances, that, at least in what is called an elementary sense, anything can be found anywhere. Mahogany logs on the coast of Greenland; bugs of a valley on the top of Mt. Blanc; atheists at a prayer meeting; ice in India. For instance, chemical analysis can reveal that almost any dead man was poisoned with arsenic, we'll say, because there is no stomach without some iron, lead, tin, gold, arsenic in it and of it—which, of course, in a broader sense, doesn't matter much, because a certain number of persons must, as a restraining influence be executed for murder every year; and, if detectives aren't able really to detect anything, illusion of their success is all that is necess-ary, and it is very honourable to give up one's life for society as a whole.

The chemist who analysed the substance of Pictou sent a sample to the Editor of the *Journal*. The Editor of course found some pollen in it.

My own acceptance is that there'd have to be some pollen in it: that nothing could very well fall through the air, in June, near the pine forests of Nova Scotia, and escape all floating spores of pollen. But the Editor does not say that this substance "contained" pollen. He disregards "Nitro-gen ammonia, and an animal odour", and says that the substance was pollen. For the sake of our thirty or forty tokens of liberality, or pseudo-liberality, if we can't be really liberal we grant that the chemist of the first examination probably wouldn't know an animal odour if he were janitor of a menagerie. As we go along, however, there can be no such sweeping ignoring of this phenomenon:

The fall of animal-matter from the sky.

I'd suggest, to start with, that we'd put ourselves in the place of deep-sea fishes:

How would they account for the fall of animal-matter from above?

They wouldn't try—

Or it's easy enough to think of most of us as deep-sea fishes of a kind.

Jour. Franklin Inst., 90–11:

That, upon the 14th of February, 1870, there fell, at Genoa, Italy according to Director Boccardo, of the Technical Institute of Genoa, and Prof. Castellani, a yellow substance. But the microscope revealed numerous globules of cobalt blue, also corpuscles of a pearly colour that resembled starch. See *Nature*, 2–166.

Comptes Rendus, 56–972:

M. Bouis says of a substance, reddish varying to yellowish, that fell enormously and successively, or upon April 30, May 1 and May 2, in France and Spain, that it carbonized and spread the odour of charred animal matter—that it was not pollen—that in alcohol it left a residue of resinous matter.

Hundreds of thousands of tons of this matter must have fallen.

"Odour of charred animal matter."

Or an aerial battle that occurred in inter-planetary space several hundred years ago—effect of time in making diverse remains uniform in appearance—

It's all very absurd because, even though we are told of a prodigious quantity of animal matter that fell from the sky—three days—France and Spain—we're not ready yet: that's all. M. Bouis says that this substance was not pollen; the vastness of the fall makes acceptable that it was not pollen; still, the resinous residue does suggest pollen of pine trees. We shall hear a great deal of a substance with a resinous residue that has fallen from the sky: finally we shall divorce it from all suggestions of pollen.

Blackwood's Magazine, 3–338:

A yellow powder that fell at Gerace, Calabria, March 14, 1813. Some of this substance was collected by Sig. Simenini, Professor of Chemistry, at Naples. It had an earthy, insipid taste, and is described as "unctuous'. When heated, this matter turned brown, then black, then red. According to the *Annals of Philosophy*, 11–466, one of the components was a greenish-yellow substance, which, when dried, was found to be resinous.

But concomitants of this fall:

Loud noises were heard in the sky.

Stones fell from the sky.

According to Chladni, these concomitants occurred, and to me they seem—rather brutal?—or not associable with something so soft and gentle as a fall of pollen?

Black rains and black snows—rains as black as a deluge of ink—jet-black snowflakes.

Such a rain as that which fell in Ireland, May 14, 1849, described in the *Annals of Scientific Discovery*, 1850, and the *Annual Register*, 1849. It fell upon a district of 400 square miles, and was the colour of ink, and of a fetid odour and very disagreeable taste.

The rain at Castlecommon, Ireland, April 30, 1887—"thick, black rain". (*Amer. Met. Jour.*, 4–193.)

A black rain fell in Ireland, Oct. 8 and 9, 1907. (*Symons' Met. Mag.*, 43–2.) "It left a most peculiar and disagreeable smell in the air."

The orthodox explanation of this rain occurs in *Nature*, March 2, 1908—cloud of soot that had come from South Wales, crossing the Irish Channel and all of Ireland.

So the black rain of Ireland, of March, 1898: ascribed in *Symons' Met Mag.*, 33–40, to clouds of soot from the manufacturing towns of North England and South Scotland.

Our Intermediatist principle of pseudo-logic, or our principle of Continuity is, of course, that nothing is unique or individual: that all phenomena merge away into all other phenomena: that, for instance—suppose there should be vast celestial super-oceanic, or inter-planetary vessels that come near this earth and discharge volumes of smoke at times. We're only supposing such a thing as that now, because, conventionally, we are beginning modestly and tentatively. But if it were so, there would necessarily be some phenomenon upon this earth, with which that phenomenon would merge. Extra-mundane smoke and smoke from cities merge, or both would manifest in black precipitations in rain.

In Continuity, it is impossible to distinguish phenomena at their merging-points, so we look for them at their extremes. Impossible to distinguish between animal and vegetable in some infusoria—but hippopotamus and violet. For all practical purposes they're distinguishable enough. No one but a Barnum or a Bailey would send one a bunch of hippopotami as a token of regard.

So away from the great manufacturing centres:

Black rain in Switzerland, Jan. 20, 1911. Switzerland is so remote, and so ill at ease is the conventional explanation here, that *Nature*, 85–451, says of this rain that in certain conditions of weather, snow may take on an appearance of blackness that is quite deceptive.

Maybe so. Or at night, if dark enough, snow may look black. This is simply denying that a black rain fell in Switzerland, Jan. 20, 1911.

Extreme remoteness from great manufacturing centres: *La Nature*, 1888, 2–406:

That Aug. 14, 1888, there fell at the Cape of Good Hope, a rain so black as to be described as a "shower of ink".

Continuity dogs us. Continuity rules us and pulls us back. We seemed to have little hope that by the method of extremes we could get away from things that merge indistinguishably into other things. We find that every departure from one merger is entrance upon another. At the Cape of Good Hope, vast volumes of smoke from great manufacturing centres, as an explanation, cannot very acceptably merge with the explanation of extra-mundane origin—but smoke from a terrestrial volcano can, and that is the suggestion that is made in *La Nature*.

There is, in human intellection, no real standard to judge by, but our acceptance, for the present, is that the more nearly positive will prevail. By the more nearly positive we mean the more nearly Organized. Everything merges away into everything else, but proportionately to its complexity, if unified, a thing seems strong, real, and distinct: so, in aesthetics, it is recognized that diversity in unity is higher beauty or approximation to Beauty, than is simpler unity; so the logicians feel that agreement of diverse data constitute greater convincingness, or strength, than that of mere parallel instances: so to Herbert Spencer the more highly differentiated and integrated is the more fully evolved. Our opponents hold out for mundane origin of all black rains. Our method will be the presenting of diverse phenomena in agreement with the notion of some other origin. We take up not only black rains but black rains and their accompanying phenomena.

A correspondent to *Knowledge*, 5–190, writes of a black rain that fell in the Clyde Valley, March 1, 1884: of another black rain that fell two days later. According to the correspondent, a black rain had fallen in the Clyde Valley, March 20, 1828: then again March 22, 1828. According to *Nature*, 9–43, a black rain fell at Marlsford, England, Sept. 4, 1873; more than twenty-four hours later another black rain fell in the same small town.

The black rains of Slains:

According to Rev. James Rust (*Scottish Showers*):

A black rain at Slains, Jan. 14, 1862—another at Carluke, 140 miles from Slains, May 1, 1862—at Slains, May 20, 1862—Slains, Oct. 28, 1863.

But after two of these showers, vast quantities of a substance described sometimes as "pumice stone", but sometimes as "slag", were washed upon the sea coast near Slains. A Chemist's opinion is given that this substance was slag: that it was not a volcanic product: slag from smelting works. We now have, for black rains, a concomitant that is irreconcilable with origin from factory chimneys. Whatever it may have been the quantity of this substance was so enormous that, in Mr. Rust's opinion, to have produced so much of it would have required the united output of all the smelting works in the world. If slag it were, we accept that an artificial product has, in enormous quantities, fallen from the sky. If you don't think that such occurrences are damned by Science, read *Scottish Showers* and see how impossible it was for the author to have this matter taken up by the scientific world.

The first and second rains corresponded, in time, with ordinary ebullitions of Vesuvius.

The third and fourth, according to Mr. Rust, corresponded with no known volcanic activities upon this earth.

La Science Pour Tous, 11–26:

That, between October, 1863, and January, 1866, four more black rains fell at Slains, Scotland.

The writer of this supplementary account tells us, with a better, or more unscrupulous, orthodoxy than Mr. Rust's, that of the eight black rains, five coincided with eruptions of Vesuvius and three with eruptions of Etna.

The fate of all explanation is to close one door only to have another fly wide open. I should say that my own notions upon this subject will be considered irrational, but at least my gregariousness is satisfied in associating here with the preposterous—or this writer, and those who think in his rut, have to say that they can think of four discharges from one far-distant volcano, passing over a great part of Europe, precipitating nowhere else, discharging precisely over one small northern parish—

But also of three other discharges, from another far-distant volcano, showing the same precise preference, if not marksmanship, for one small parish in Scotland.

Nor would orthodoxy be any better off in thinking of exploding meteorites and their débris: preciseness and recurrence would be just as difficult to explain.

My own notion is of an island near an oceanic trade-route: it might receive débris from passing vessels seven times in four years.

Other concomitants of black rains:

In Timb's *Year Book*, 1851–270, there is an account of "a sort of rumbling, as of wagons, heard for upward of an hour without ceasing", July 16, 1850, Bulwick Rectory, Northampton, England. On the 19th, a black rain fell.

In *Nature*, 30–6, a correspondent writes of an intense darkness at Preston, England, April 26, 1884: page 32, another correspondent writes of black rain at Crowle, near Worcester, April 26: that a week later, on May 3, it had fallen again: another account of black rain, upon the 28th of April, near Church Stretton, so intense that the following brooks were still dyed with it. According to four accounts by correspondents to *Nature* there were earthquakes in England at this time.

Or the black rain of Canada, Nov. 9, 1819. This time it is orthodoxy to attribute the black precipitate to smoke of forest fires south of the Ohio River—

Zurcher, *Meteors*, p, 238:

That this black rain was accompanied by "shocks like those of an earthquake".

43

Edinburgh Philosophical Journal, 2–381:
That the earthquake had occurred at the climax of intense darkness and the fall of black rain.

Red rains.

Orthodoxy:

Sand blown by the sirocco, from the Sahara to Europe.

Especially in the earthquake regions of Europe, there have been many falls of red substance, usually, but not always precipitated in rain. Upon many occasions, these substances have been "absolutely identified" as sand from the Sahara. When I first took this matter up, I came across assurance after assurance, so positive to this effect, that, had I not been an Intermediatist, I'd have looked no farther. Samples collected from rain at Genoa—samples of sand forwarded from the Sahara—"absolute agreement" some writers said: same colour, same particles of quartz, even the same shells of diatoms mixed in. Then the chemical analyses: not a disagreement worth mentioning.

Our intermediatist means of expression will be that, with proper exclusions, after the scientific or theological method, anything can be identified with anything else if all things are only different expressions of an underlying oneness.

To many minds there's rest and there's satisfaction in that expression "absolutely identified". Absoluteness or the illusion of it—the universal quest. If chemists have identified substances that have fallen in Europe as sand from African deserts, swept up in African whirlwinds, that's assuasive to all the irritations that occur to those cloistered minds that must repose in the concept of a snug, isolated, little world, free from contact with cosmic wickednesses, safe from stellar guile, undisturbed by interplanetary prowlings and invasions. The only trouble is that a chemist's analysis, which seems so final and authoritative to some minds, is no more nearly absolute than is identification by a child or description by an imbecile—

I take some of that back: I accept that the approximation is higher—

But that it's based upon delusion, because there is no definiteness, no homogeneity, no stability, only different stages somewhere between them and indefiniteness, heterogeneity, and instability. There are no chemical ele-

44

ments. It seems acceptable that Ramsey and others have settled that. The chemical elements are only another disappointment in the quest for the positive, as the definite, the homogeneous, and the stable. If there were real elements, there could be a real science of chemistry.

Upon Nov. 12 and 13, 1902, occurred the greatest fall of matter in the history of Australia. Upon the 14th of November, it "rained mud", in Tasmania. It was of course attributed to Australian whirlwinds, but, according to the *Monthly Weather Review*, 32–365, there was a haze all the way to the Philippines, also as far as Hong Kong. It may be that this phenomenon had no especial relation with the even more tremendous fall of matter that occurred in Europe, February, 1903.

For several days, the south of England was a dumping ground—from somewhere.

If you'd like to have a chemist's opinion, even though it's only a chemist's opinion, see the report of the meeting of the Royal Chemical Society, April 2, 1903. Mr. E. G. Clayton read a paper upon some of the substance that had fallen from the sky, collected by him. The Sahara explanation applies most to falls that occur in southern Europe. Farther away, the conventionalists are a little uneasy: for instance, the editor of the *Monthly Weather Review*, 29–121, says of the red rain that fell near the coast of Newfoundland, early in 1890: "It would be very remarkable if this was Sahara dust." Mr. Clayton said that the matter examined by him was "merely wind-borne dust from the roads and lanes of Wessex". This opinion is typical of all scientific opinion—or theological opinion—or feminine opinion—all very well except for what it disregards. The most charitable thing I can think of—because I think it gives us a broader tone to relieve our malices with occasional charities—is that Mr. Clayton had not heard of the astonishing extent of this fall—had covered the Canary Islands, on the 19th, for instance. I think, myself, that in 1903, we passed through the remains of a powdered world—left over from an ancient inter-planetary dispute, brooding in space like a red resentment ever since. Or, like every other opinion, the notion of dust from Wessex turns into a provincial thing when we look it over.

To think is to conceive incompletely, because all thought relates only to the local. We metaphysicians, of course, like to have the notion that we think of the unthinkable.

As to opinions, or pronouncements, I should say, because they always have such an authoritative air, of other chemists, there is an analysis in *Nature*, 68–54, giving water and organic matter at 9.08 per cent. It's that carrying out of fractions that's so convincing. The substance is identified as sand from the Sahara.

The vastness of this fall. In *Nature*, 68–65, we are told that it had occurred in Ireland, too. The Sahara, of course—because prior to February 19, there had been dust storms in the Sahara—disregarding that in that great region there's always, in some part of it, a dust storm. However, just at present, it does look reasonable that dust had come from Africa, via the Canaries.

The great difficulty that authoritativeness has to contend with is some other authoritativeness. When an infallibility clashes with a pontification—

They explain.

Nature, March 5, 1903:

Another analysis—36 per cent organic matter.

Such disagreements don't look very well, so, in *Nature*, 68–109, one of the differing chemists explains. He says that his analysis was of muddy rain, and the other was of sediment of rain—

We're quite ready to accept excuses from the most high, though I do wonder whether we're quite so damned as we were, if we find ourselves in a gracious and tolerant mood toward the powers that condemn—but the tax that now comes upon our good manners and unwillingness to be too severe—

Nature, 68–223:

Another chemist. He says it was 23.49 per cent water and organic matter.

He "identifies" this matter as sand from an African desert—but after deducting organic matter—

But you and I could be "identified" as sand from an African desert, after deducting all there is to us except sand—

Why we cannot accept that this fall was of sand from the Sahara, omitting the obvious objection that in most parts the Sahara is not red at all, but is usually described as "dazzling white"—

The enormousness of it: that a whirlwind might have carried it, but that, in that case it would be no suppositious, or doubtfully identified whirlwind, but the greatest atmospheric cataclysm in the history of this earth:

Jour. Roy. Met. Soc., 30–56:

That, up to the 27th of February, this fall had continued in Belgium, Holland, Germany and Austria; that in some instances it was not sand, or that almost all the matter was organic: that a vessel had reported the fall as occurring in the Atlantic Ocean, midway between Southampton and the Barbados. The calculation is given that, in England alone, 10,000,000 tons of matter had fallen. It had fallen in Switzerland (*Symons' Met. Mag.*, March, 1903). It had fallen in Russia (*Bull. Com. Geolog.*, 22–48). Not only had a vast quantity of matter fallen several months before, in Australia, but it was at this time falling in Australia, (*Victorian Naturalist*, June, 1903)—enormously—red mud—fifty tons per square mile.

The Wessex explanation—

Or that every explanation is a Wessex explanation: by that I mean an attempt to interpret the enormous in terms of the minute—but that nothing can be finally explained, because by Truth we mean the Universal; and that even if we could think as wide as Universality, that would not be requital to the cosmic quest—which is not for Truth, but for the local that is true—not to universalize the local, but to localize the universal—or to give to a cosmic cloud absolute interpretation in terms of the little dusty roads and lanes of Wessex. I cannot conceive that this can be done: think of high approximation.

Our Intermediatist concept is that, because of the continuity of all "things", which are not separate, positive, or real things, all pseudo-things partake of the underlying, or are only different expressions, degrees, or aspects of the underlying; so then that a sample from somewhere in anything must correspond with a sample from somewhere in anything else.

That, by due care in selection, and disregard for everything else, or the scientific and theological method, the substance that fell, February, 1903, could be identified with anything, or with some part or aspect of anything that could be conceived of—

With sand from the Sahara, sand from a barrel of sugar,

or dust of your great-great-grandfather.

Different samples are described and listed in the *Journal of the Royal Meteorological Society*, 30–57—or we'll see whether my notion that a chemist could have identified some one of these samples as from anywhere conceivable, is extreme or not:

"Similar to brick dust", in one place; "buff or light brown", in another place; "chocolate-coloured and silky to the touch and slightly iridescent"; "grey"; "red-dust colour"; "reddish raindrops and grey sand"; "dirty grey"; "quite red"; "yellow-brown, with a tinge of pink"; "deep yellow-clay colour".

In Nature, it is described as of a peculiar yellowish cast in one place, reddish somewhere else, and salmon-coloured in another place.

Or there could be real science if there were really anything to be scientific about.

Or the science of chemistry is like a science of sociology, prejudiced in advance, because only to see is to see with a prejudice, setting out to "prove" that all inhabitants of New York came from Africa.

Very easy matter. Samples from one part of town. Disregard for all the rest.

There is no science but Wessex-science.

According to our acceptance, there should be no other, but that approximation should be higher: that metaphysics is super-evil: that the scientific spirit is of the cosmic quest.

Our notion is that, in a real existence, such a quasi-system of fables as the science of chemistry could not deceive for a moment: but that in an "existence" endeavouring to become real, it represents that endeavour, and will continue to impose its pseudo-positiveness until it be driven out by a higher approximation to realness:

That the science of chemistry is as impositive as fortune-telling—

Or no—

That, though it represents a higher approximation to realness than does alchemy, for instance, and so drove out alchemy, it is still only somewhere between myth and positiveness.

The attempt at realness, or to state a real and unmodified fact here, is the statement:

All red rains are coloured by sands from the Sahara

Desert.

My own impositivist acceptances are:

That some red rains are coloured by sands from the Sahara Desert;

Some by sands from other terrestrial sources:

Some by sands from other worlds, or from their deserts—also from aerial regions too indefinite or amorphous to be thought of as "worlds" or planets—

That no supposititious whirlwind can account for the hundreds of millions of tons of matter that fell upon Australia, Pacific Ocean and Atlantic Ocean and Europe in 1902 and 1903—that a whirlwind that could do that would not be supposititious.

But now we shall cast off some of our own wessicality by accepting that there have been falls of red substance other than sand.

We regard every science as an expression of the attempt to be real. But to be real is to localize the universal—or to make some one thing as wide as all things—successful accomplishment of which I cannot conceive. The prime resistance to this endeavour is the refusal of the rest of the universe to be damned, excluded, disregarded, to receive Christian Science treatment, by something else so attempting. Although all phenomena are striving for the Absolute—or have surrendered to and have incorporated themselves in higher attempts, simply to be phenomenal, or to have seeming in Intermediateness is to express relations.

A river.

It is water expressing the gravitational relations of different levels.

The water of the river.

Expression of chemic relations of hydrogen and oxygen—which are not final.

A city.

Manifestation of commercial and social relations.

How could a mountain be without base in a greater body, Storekeeper live without customers?

The prime resistance to the positivist attempt by Science is its relations with other phenomena, or that it only expresses those relations in the first place. Or that a Science can have seeming, or survive in Intermediateness, as something pure, isolated, positively different, no more than could a river or a city or a mountain or a store.

49

This Intermediateness-wide attempt by parts to be wholes—which cannot be realized in our quasi-state, if we accept that in it the co-existence of two or more wholes or universals is impossible—high approximation to which, however, may be thinkable—

Scientists and their dream of "pure science".

Artists and their dream of "art for art's sake".

It is our notion that if they could almost realize, that would be almost realness: that they would instantly be translated into real existence. Such thinkers are good positivists, but they are evil in an economic and sociologic sense, if, in that sense, nothing has justification for being, unless it serve, or function for, or express the relations of, some higher aggregate. So Science functions for and serves society at large, and would, from society at large, receive no support, unless it did so divert itself or dissipate and prostitute itself. It seems that by prostitution I mean usefulness.

There have been red rains that, in the middle ages, were called "rains of blood". Such rains terrified many persons, and were so unsettling to large populations, that Science, in its sociologic relations, has sought, by Mrs. Eddy's method, to remove an evil—

That "rains of blood" do not exist;

That rains so called are only of water coloured by sand from the Sahara Desert.

My own acceptance is that such assurances, whether fictitious or not, whether the Sahara is a "dazzling white" desert or not, have wrought such good effects, in a sociologic sense, even though prostitutional in the positivist sense, that, in the sociologic sense, they were well justified;

But that we've gone on: that this is the twentieth century; that most of us have grown up so that such soporifics of the past are no longer necessary:

That if gushes of blood should fall from the sky upon New York, business would go on as usual.

We begin with rains that we accepted ourselves were, most likely, only of sand. In my own still immature hereticalness—and by heresy, or progress, I mean, very largely, a return, though with many modifications, to the superstitions of the past, I think I feel considerable aloofness to the idea of rains of blood. Just at present, it is my conservative, or timid purpose, to express only that there have been red rains that very strongly suggest blood or finely divided animal matter—

Débris from inter-planetary disasters.

Aerial battles.

Food-supplies from cargoes of super-vessels, wrecked in inter-planetary traffic.

There was a red rain in the Mediterranean region, March 6, 1888. Twelve days later, it fell again. Whatever this substance may have been, when burned, the odour of animal matter from it was strong and persistent. (*L'Astronomie*, 1888–205.)

But—infinite heterogeneity—or débris from many different kinds of aerial cargoes—there have been red rains that have been coloured by neither sand nor animal matter.

Annals of Philosophy, 16–226:

That, Nov. 2, 1819—week before the black rain and earthquake of Canada—there fell, at Blankenberge, Holland, a red rain. As to sand, two chemists of Bruges concentrated 144 ounces of the rain to 4 ounces—"no precipitate fell". But the colour was so marked that had there been sand, it would have been deposited, if the substance had been diluted instead of concentrated. Experiments were made, and various reagents did cast precipitates, but other than sand. The chemists concluded that the rain-water contained muriate of cobalt—which is not very enlightening: that could be said of many substances carried in vessels upon the Atlantic Ocean. Whatever it may have been, in the *Annals de Chimie*, 2–12–432, its colour is said to have been red-violet. For various chemic reactions, see *Quar. Jour. Roy. Inst.*, 9–202, and *Edin. Phil. Jour.*, 2–381.

Something that fell with dust said to have been meteoric, March 9, 10, 11, 1872: described in the *Chemical News*, 25–300, as a "peculiar substance", consisted of red iron ochre, carbonate of lime, and organic matter.

Orange, red hail, March 14, 1873, in Tuscany. (*Notes and Queries* 9–5–16.)

Rain of lavender-coloured substance, at Oudon, France, Dec. 19, 1903. (*Bull. Soc. Met. de France*, 1904–124.)

La Nature, 1885–2–351:

That, according to Prof. Schwedoff, there fell, in Russia, June 14, 1880, red hailstones, also blue hailstones, also grey hailstones.

Nature, 34–123:

A correspondent writes that he had been told by a resident of a small town in Venezuela, that there, April 17, 1886, had fallen hailstones, some red, some blue, some whitish: informant said to have been one unlikely ever to have heard of the Russian phenomenon; described as an "honest, plain countryman".

Nature, July 5, 1877, quotes a Roman correspondent to the London *Times* who sent a translation from an Italian newpaper: that a red rain had fallen in Italy, June 23, 1887, containing "microscopically small particles of sand".

Or, according to our acceptance, any other story would have been an evil thing, in the sociologic sense, in Italy, in 1877. But the English correspondent, from a land where terrifying red rains are uncommon, does not feel this necessity. He writes: "I am by no means satisfied that the rain was of sand and water." His observations are that drops of this rain left stains "such as sandy water could not leave". He notes that when the water evaporated, no sand was left behind.

L'Année Scientifique, 1888–75:

That, Dec. 13, 1887, there fell, in Cochin China, a substance like blood, somewhat coagulated.

Annales de Chimie, 85–266:

That a thick, viscous, red matter fell at Ulm, in 1912.

We now have a datum with a factor that has been foreshadowed; which will recur and recur and recur throughout this book. It is a factor that makes for speculation so revolutionary that it will have to be reinforced many times before we can take it into full acceptance.

Year Book of Facts, 1861–273:

Quotation from a letter from Prof. Campini to Prof. Matteucci:

That, upon Dec. 28, 1860, at about 7 A.M., in the north-western part of Siena, a reddish rain fell copiously for two hours.

A second red shower fell at 11 o'clock.

Three days later, the rain fell again.

The next day another red rain fell.

Still more extraordinarily:

Each fall occurred in "exactly the same quarter of town".

Chapter 4

It is in the records of the French Academy that, upon March 17, 1669, in the town of Châtillon-sur-seine, fell a reddish substance that was "thick, viscous, and putrid".

American Journal of Science, 1–41–404:

Story of a highly unpleasant substance that had fallen from the sky, in Wilson County, Tennessee. We read that Dr. Troost visited the place and investigated. Later we're going to investigate some investigations—but never mind that now. Dr. Troost reported that the substance was clear blood and portions of flesh scattered upon tobacco fields. He argued that a whirlwind might have taken an animal up from one place, mauled it around, and have precipitated its remains somewhere else.

But, in volume 44, page 216, of the *Journal*, there is an apology. The whole matter is, upon newspaper authority, said to have been a hoax by Negroes, who had pretended to have seen the shower, for the sake of practising upon the credulity of their masters: that they had scattered the decaying flesh of a dead hog over the tobacco fields.

If we don't accept this datum, at least we see the sociologically necessary determination to have all falls accredited to earthly origins—even when they're falls that don't fall.

Annual Register, 1821–687:

That, upon the 13th of August, 1819, something had fallen from the sky at Amherst, Mass. It had been examined and described by Prof. Graves, formerly lecturer at Dartmouth College. It was an object that had upon it a nap, similar to that of milled cloth. Upon removing this nap, a buff-coloured, pulpy substance was found. It had an offensive odour, and, upon exposure to the air, turned to a vivid red. This thing was said to have fallen with a brilliant light.

Also see the *Edinburgh Philosophical Journal*, 5–295. In the *Annales de Chimie*, 1821–67, M. Arago accepts the datum, and gives four instances of similar objects or substances said to have fallen from the sky, two of which we shall have with our data of gelatinous, or viscous matter, and two of which I omit, because it seems to me that the dates given are too far back.

In the *American Journal of Science*, 1–2–335, is Professor Graves' account, communicated by Professor Dewey:

That, upon the evening of August 13, 1819, a light was seen in Amherst—a falling object—sound as if of an explosion.

In the home of Prof. Dewey, this light was reflected upon a wall of a room in which were several members of Prof. Dewey's family.

The next morning, in Prof. Dewey's front yard, in what is said to have been the only position from which the light that had been seen in the room, the night before, could have been reflected, was found a substance "unlike anything before observed by anyone who saw it". It was a bowl-shaped object, about eight inches in diameter, and one inch thick. Bright buff-coloured, and having upon it a "fine nap". Upon removing this covering, a buff-coloured, pulpy substance of the consistency of soft-soap, was found "of an offensive, suffocating smell".

A few minutes of exposure to the air changed the buff colour to "a livid colour resembling venous blood". It absorbed moisture quickly from the air and liquefied. For some of the chemic reactions, see the *Journal*.

There's another lost quasi-soul of a datum that seems to me to belong here:

London *Times*, April 19, 1836:

Fall of fish that had occurred in the neighbourhood of Allahabad, India. It is said that the fish were of the chalwa species, about a span in length and a seer in weight—you know.

They were dead and dry.

Or they had been such a long time out of water that we can't accept that they had been scooped out of a pond, by a whirlwind—even though they were so definitely identified as of a known local species—

Or they were not fish at all.

I incline, myself, to the acceptance that they were not fish, but slender, fish-shaped objects of the same substance as that which fell at Amherst—it is said that, whatever they were, they could not be eaten: that "in the pan, they turned to blood".

For details of this story see the *Journal of the Asiatic*

Society of Bengal, 1834–307. May 16 or 17, 1834, is the date given in the *Journal*.

In the *American Journal of Science*, 1–25–362, occurs the inevitable damnation of the Amherst object:

Prof. Edward Hitchcock went to live in Amherst. He says that years later, another object, like the one said to have fallen in 1819, had been found at "nearly the same place". Prof. Hitchcock was invited by Prof. Graves to examine it. Exactly like the first one. Corresponded in size and colour and consistency. The chemic reactions were the same.

Prof. Hitchcock recognized it in a moment.

It was a gelatinous fungus.

He did not satisfy himself as to just the exact species it belonged to, but he predicted that similar fungi might spring up within twenty-four hours—

But, before evening, two others sprang up.

Or we've arrived at one of the oldest of the exclusionists' conventions—or nostoc. We shall have many data of gelatinous substance said to have fallen from the sky: almost always the exclusionists argue that it was only nostoc, an Alga, or, in some respects, a fungous growth. The rival convention is "spawn of frogs or of fishes". These two conventions have made a strong combination. In instances where testimony was not convincing that gelatinous matter had been seen to fall, it was said that the gelatinous substance was nostoc, and had been upon the ground in the first place: when the testimony was too good that it had fallen, it was said to be spawn that had been carried from one place to another in a whirlwind.

Now, I can't say that nostoc is always greenish, any more than I can say that blackbirds are always black, having seen a white one: we shall quote a scientist who knew of flesh-coloured nostoc, when so to know was convenient. When we come to reported falls of gelatinous substances, I'd like it to be noticed how often they are described as whitish or greyish. In looking up the subject, myself, I have read only of greenish nostoc. Said to be greenish, in Webster's Dictionary—said to be "blue-green" in the New International Encyclopedia—"from bright green to olive-green" (*Science Gossip*, 10–114); "green" (*Science Gossip*, 7–260); "greenish" (*Notes and Queries*, 1–11–219). It would seem acceptable that, if many reports of white birds should occur, the birds are not blackbirds, even though there have been white blackbirds. Or that, if often reported, greyish or whitish gelatinous substance is not nostoc, and is not spawn if occurring in times unseasonable for spawn.

"The Kentucky Phenomenon."

So it was called, in its day, and now we have an occurrence that attracted a great deal of attention in its own time. Usually these things of the accursed have been hushed up or disregarded—suppressed like the seven black rains of Slains—but, upon March 3, 1876, something occurred in Bath County, Kentucky, that brought many newspaper correspondents to the scene.

The substance that looked like beef that fell from the sky.

Upon March 3, 1876, at Olympian Springs, Bath County, Kentucky, flakes of a substance that looked like beef fell from the sky—"from a clear sky". We'd like to emphasize that it was said that nothing but this falling substance was visible in the sky. It fell in flakes of various sizes; some two inches square, one, three or four inches square. The flake-formation is interesting: later we shall think of it as signifying pressure—somewhere. It was a thick shower, on the ground, on trees, on fences, but it was narrowly localized: or upon a strip of land about 100 yards long and about 50 yards wide. For the first account, see the *Scientific American*, 34–197 and the *New York Times*, March 10, 1876.

Then the exclusionists.

Something that looked like beef: one flake of it the size of a square envelope.

If we think of how hard the exclusionists have fought to reject the coming of ordinary-looking dust from this earth's externality, we can sympathize with them in this sensational instance, perhaps. Newspaper correspondents wrote broadcast and witnesses were quoted, and this time there is no mention of a hoax, and, except by one scientist, there is no denial that the fall did take place.

It seems to me that the exclusionists are still more emphatically conservators. It is not so much that they are inimical to all data of externally derived substances that fall upon this earth, as that they are inimical to all data discordant with a system that does not include such phenomena—

Or the spirit or hope or ambition of the cosmos, which we call attempted positivism: not to find out the new; not to

56

add to what is called knowledge, but to systematize.

Scientific American Supplement, 2–426:

That the substance reported from Kentucky had been examined by Leopold Brandeis.

"At last we have a proper explanation of this much talked of phenomenon."

"It has been comparatively easy to identify the substance and to fix its status. The Kentucky 'wonder' is no more or less than nostoc."

Or that it had not fallen; that it had been upon the ground in the first place, and had swollen with rain, and, attracting attention by greatly increased volume, had been supposed by unscientific observers to have fallen in rain—

What rain, I don't know.

Also it is spoken of as "dried" several times. That's one of the most important of the details.

But the relief of outraged propriety, expressed in the *Supplement*, is amusing to some of us, who, I fear, may be a little improper at times. Very spirit of the Salvation Army, when some third-rate scientist comes out with an explanation of the vermiform appendix or the os coccyges that would have been acceptable to Moses. To give completeness to "the proper explanation", it is said the Mr. Brandeis had identified the substance as "flesh-coloured" nostoc.

Prof. Lawrence Smith, of Kentucky, one of the most resolute of the exclusionists:

New York Times, March 12, 1876:

That the substance had been examined and analysed by Prof. Smith, according to whom it gave every indication of being the "dried" spawn of some reptile, "doubtless of the frog"—or up from one place and down in another. As to "dried", that may refer to condition when Prof. Smith received it.

In the *Scientific American Supplement*, 2–473, Dr. A. Mead Edwards, President of the Newark Scientific Association, writes that, when he saw Mr. Brandeis' communication, his feeling was of conviction that propriety had been re-established, or that the problem had been solved, as he expressed it: knowing Mr. Brandeis well, he had called upon that upholder of respectability, to see the substance that had been identified as nostoc. But he had also called upon Dr. Hamilton, who had a specimen, and Dr. Hamilton had declared it to be lung-tissue. Dr. Edwards writes of the substance that had so completely or beautifully—if beauty is completeness—been identified as nostoc—"It turned out to be lung tissue also." He wrote to other persons who had specimens, and identified other specimens as masses of cartilage or muscular fibres. "As to whence it came, I have no theory." Nevertheless he endorses the local explanation—and a bizarre thing it is:

A flock of gorged, heavy-weighted buzzards, but far up and invisible in the clear sky—

They had disgorged.

Prof. Fassig lists the substance, in his "Bibliography," as fish spawn. McAtee (*Monthly Weather Review*, May, 1918) lists it as a jelly-like material, supposed to have been the "dried" spawn either of fishes or of some batrachian.

Or this is why, against the seemingly insuperable odds against all things new, there can be what is called progress—

That nothing is positive, in the aspects of homogeneity and unity:

If the whole world should seem to combine against you, it is only unreal combination, or intermediateness to unity and disunity. Every resistance is itself divided into parts resisting one another. The simplest strategy seems to be—never bother to fight a thing: set its own parts fighting one another.

We are merging away from carnal to gelatinous substance, and here there is an abundance of instances or reports of instances. These data are so improper they're obscene to the science of today, but we shall see that science, before it became so rigorous, was not so prudish. Chladni was not, and Greg was not.

I shall have to accept, myself, that gelatinous substance has often fallen from the sky—

Or that, far up, or far away, the whole sky is gelatinous?

That meteors tear through and detach fragments?

That fragments are brought down by storms?

That the twinkling of stars is penetration of light through something that quivers?

I think, myself, that it would be absurd to say that the whole sky is gelatinous: it seems more acceptable that only certain areas are.

Humboldt (*Cosmos*, 1–119) says that all our data in this

respect must be "classed amongst the mythical fables of mythology". He is very sure, but just a little redundant.

We shall be opposed by the standard resistances:

There in the first place.

Up from one place, in a whirlwind, and down in another.

We shall not bother to be very convincing one way or another, because of the over-shadowing of the datum with which we shall end up. It will mean that something had been in a stationary position for several days over a small part of a town in England: that is the revolutionary thing that we have alluded to before; whether the substance were nostoc, or spawn, or some kind of a larval nexus, doesn't matter so much. If it stood in the sky for several days, we rank with Moses as a chronicler of improprieties—or was that story, or datum, we mean, told by Moses? Then we shall have so many records of gelatinous substance said to have fallen with meteorites, that, between the two phenomena, some of us will have to accept connection—or that there are at least vast gelatinous areas aloft, and that meteorites tear through, carrying down some of the substance.

Comptes Rendus, 3–554:

That, in 1836, M. Vallot, member of the French Academy, placed before the Academy some fragments of a gelatinous substance, said to have fallen from the sky, and asked that they be analysed. There is no further allusion to this subject.

Comptes Rendus, 23–542:

That, in Wilna, Lithuania, April 4, 1846, in a rainstorm, fell nut-sized masses of a substance that is described as both resinous and gelatinous. It was odourless until burned: then it spread a very pronounced sweetish odour. It is described as like gelatine, but much firmer: but, having been in water 24 hours, it swelled out, and looked altogether gelatinous—

It was greyish.

We are told that, in 1841 and 1846, a similar substance had fallen in Asia Minor.

In *Notes and Queries*, 8–6–190, it is said that, early in August, 1894, thousands of jellyfish, about the size of a shilling, had fallen at Bath, England. I think it is not acceptable that they were jellyfish: but it does look as if this time frog spawn did fall from the sky, and may have been translated by a whirlwind—because, at the same time, small frogs fell at Wigan, England.

Nature, 87–10:

That, June 24, 1911, at Eton, Bucks, England, the ground was found covered with masses of jelly, the size of peas, after a heavy rainfall. We are not told of nostoc this time: it is said that the object contained numerous eggs of "some species of Chironomus, from which larvae soon emerged".

I incline, then, to think that the objects that fell at Bath were neither jellyfish nor masses of frog spawn, but something of a larval kind—

This is what had occurred at Bath, England, 23 years before.

London *Times*, April 24, 1871, a storm of glutinous drops neither jellyfish, nor masses of frog spawn, but something of a [line missing here in original text. Ed.] railroad station, at Bath. "Many soon developed into a wormlike chrysalis, about an inch in length." The account of this occurrence in the *Zoologist*, 2–6–2686, is more like the Eton-datum: of minute forms, said to have been infusoria; not forms about an inch in length.

Trans. Ent. Soc. of London, 1871-proc. xxii:

That the phenomenon has been investigated by the Rev. L. Jenyns, of Bath. His description is of minute worms in filmy envelopes. He tries to account for their segregation. The mystery of it is: What could have brought so many of them together? Many other falls we shall have record of, and in most of them segregation is the great mystery. A whirlwind seems anything but a segregative force. Segregation of things that have fallen from the sky has been avoided as most deep-dyed of the damned. Mr. Jenyns conceives of a large pool, in which were many of these spherical masses: of the pool drying up and concentrating all in a small area; of a whirlwind then scooping all up together—

But several days later, more of these objects fell in the same place.

That such marksmanship is not attributable to whirlwinds seems to me to be what we think we mean by common sense:

It may not look like common sense to say that these things had been stationary over the town of Bath, several days—

The seven black rains of Slains;

The four red rains of Siena.

An interesting sidelight on the mechanics of orthodoxy is that Mr. Jenyns dutifully records the second fall, but ignores it in his explanation.

R. P. Greg, one of the most notable of cataloguers of meteoritic phenomena, records (*Phil. Mag.*: 4–8–463) falls of viscid substance in the years 1652, 1686, 1718, 1796, 1811, 1819, 1844. He gives earlier dates, but I practise exclusions, myself. In the *Report of the British Association*, 1860–63, Greg records a meteor that seemed to pass near the ground, between Barsdorf and Freiburg, Germany: the next day a jelly-like mass was found in the snow—

Unseasonableness for either spawn or nostoc.

Greg's comment in this substance is: "curious if true." But he records without modification the fall of a meteorite at Gotha, Germany, Sept. 6, 1835, "leaving a jelly-like mass on the ground". We are told that this substance fell only three feet away from an observer. In the *Report of the British Association*, 1855–94, according to a letter from Greg to Prof. Baden-Powell, at night, Oct. 8, 1844, near Coblenz, a German, who was known to Greg, and another person saw a luminous body fall close to them. They returned next morning and found a gelatinous mass of greyish colour.

According to Chladni's account (*Annals of Philosophy*, n.s., 12–94) a viscous mass fell with a luminous meteorite between Siena and Rome, May, 1652; viscous matter found after the fall of a fire ball, in Lusatia, March, 1796; fall of a gelatinous substance, after the explosion of a meteorite, near Heidelberg, July, 1811. In the *Edinburgh Philosophical Journal* 1–234, the substance that fell at Lusatia is said to have been the "colour and odour of dried, brown varnish". In the *Amer. Jour. Sci.*, 1–26–133, it is said that gelatinous matter fell with a globe of fire, upon the island of Lethy, India, 1718.

In the *Amer. Jour. Sci.*, 1–26–396, in many observations upon the meteors of November, 1833, are reports of falls of gelatinous substance:

That, according to newpaper reports, "lumps of jelly" were found on the ground at Rahway, N.J. The substance was whitish, or resembled the coagulated white of an egg;

That Mr. H. H. Garland, of Nelson County, Virginia, had found a jelly-like substance of about the circumference of a twenty-five-cent piece;

That, according to a communication from A. C. Twining to Prof. Olmstead, a woman at West Point, N.Y., had seen a mass the size of a teacup. It looked like boiled starch;

That, according to a newspaper of Newark, N.J., a mass of gelatinous substance, like soft soap, had been found. "It possessed little elasticity, and, on the application of heat, it evaporated as readily as water."

It seems incredible that a scientist would have such hardihood, or infidelity, as to accept that these things had fallen from the sky: nevertheless, Prof. Olmstead, who collected these lost souls, says:

"The fact that the supposed deposits were so uniformly described as gelatinous substance forms a presumption in favour of the supposition that they had the origin ascribed to them."

In contemporaneous scientific publications considerable attention was given to Prof. Olmstead's series of papers upon this subject of the November meteors. You will not find one mention of the part that treats of gelatinous matter.

Chapter 5

I shall attempt not much of correlation of dates. A mathematic-minded positivist, with his delusion that in an intermediate state twice two are four, whereas, if we accept Continuity we cannot accept that there are anywhere two things to start with, would search our data for periodicities. It is so obvious to me that the mathematical, or the regular, is the attribute of the Universal, that I have not much inclination to look for it in the local. Still, in this solar system, "as a whole", there is considerable approximation to regularity; or the mathematic is so nearly localized that eclipses, for instance, can, with rather high approximation, be foretold, though I have notes that would deflate a little the astronomers' vainglory in this respect—or would if that were possible. An astronomer is poorly paid, uncheered by crowds, considerably isolated: he lives upon his own inflations: deflate a bear and it couldn't hibernate. This solar system is like every other phenomenon that can be regarded "as a whole"—or the affairs of a ward are interfered with by the affairs of the city of which it is a part; city by county; county by state; state by nation; nation by other nations; all nations by climatic conditions by solar circumstances; sun by general planetary circumstances; solar system "as a whole" by other solar systems—so the hopelessness of finding the phenomena of entirety in the ward of a city. But positivists are those who try to find the unrelated in the ward of a city. In our acceptance this is the spirit of cosmic religion. Objectively the state is not realizable in the ward of a city. But, if a positivist could bring himself to absolute belief that he had found it, that would be a subjective realization of that which is unrealizable objectively. Of course we do not draw a positive line between the objective and the subjective—or that all phenomena called things or persons are subjective within one all-inclusive nexus, and that thoughts within those that are commonly called "persons" are sub-subjective. It is rather as if Intermediateness strove for Regularity in this solar system and failed: then generated the mentality of astronomers, and, in that secondary expression strove for conviction that failure had been success.

I have tabulated all the data of this book, and a great deal besides—card system—and several proximities, thus emphasized, have been revelations to me: nevertheless, it is only the method of theologians and scientists—worst of all, of statisticians.

For instance, by the statistic method, I could "prove" that a black rain has fallen "regularly" every seven months, somewhere upon this earth. To do this, I'd have to include red rains and yellow rains, but conventionally, I'd pick out the black particles in red substances and in yellow substances, and disregard the rest. Then, too, if here and there a black rain should be a week early or a month late—that would be "acceleration" or "retardation". This is supposed to be legitimate in working out the periodicities of comets. If black rains, or red or yellow rains with black particles in them, should not appear at all near some dates—we have not read Darwin in vain—"the records are not complete". As to other, interfering black rains, they'd be either grey or brown, or for them we'd find other periodicities.

Still, I have had to notice the year 1819, for instance. I shall not note them all in this book, but I have records of 31 extraordinary events in 1883. Someone should write a book upon the phenomena of this one year—that is, if books should be written. 1849 is notable for extraordinary falls, so far apart that a local explanation seems inadequate—not only the black rain of Ireland, May, 1849, but a red rain in Sicily and a red rain in Wales. Also, it is said (Timb's *Year Book* 1850–241) that, upon April 18 or 20, 1849, shepherds near Mt. Ararat found a substance that was not indigenous, upon areas measuring 8 to 10 miles in circumference. Presumably it had fallen there.

We have already gone into the subject of Science and its attempted positiveness and its resistances in that it must have relations of service. It is very easy to see that most of the theoretic science of the nineteenth century was only a relation of reaction against theologic dogma, and has no more to do with Truth than has a wave that bounds back from a shore. Or, if a shop girl, or you or I, should pull out a piece of chewing gum about a yard long, that would be quite as scientific a performance as was the stretching of this earth's age several hundred millions of years.

All "things" are not things, but only relations, or expressions of relations: but all relations are striving to be the unrelated, or have surrendered to, and subordinated to, higher attempts. So there is a positivist aspect to this reaction that is itself only a relation, and that is the attempt to assimilate all phenomena under the materialist explanation, or to formulate a final, all-inclusive system, upon the materialist basis. If this attempt could be realized that would be the attaining of realness; but this attempt can be made only by disregarding psychic phenomena, for instance—or, if science shall eventually give in to the psychic, it would be no more legitimate to explain the immaterial in terms of the material than to explain the material in terms of the immaterial. Our own acceptance is that material and immaterial are of a oneness, merging, for instance, in a thought that is continuous with a physical action: that oneness cannot be explained, because the process of explaining is the interpreting of something in terms of something else. All explanation is assimilation of something in terms of something else that has been taken as a basis: but in Continuity, there is nothing that is any more basic than anything else—unless we think that delusion built upon delusion is less real than its pseudo-foundation.

In 1829 (Timb's *Year Book*, 1848–235) in Persia fell a substance that the people said they had never seen before. As to what it was, they had not a notion, but they saw that the sheep ate it. They ground it into flour and made bread, said to have been passable enough, though insipid.

That was a chance that science did not neglect. Manna was placed upon a reasonable basis, or was assimilated and reconciled with the system that had ousted the older—and less nearly real—system. It was said that, likely enough, manna had fallen in ancient times—because it was still falling—but that there was no tutelary influence behind it—that it was a lichen from the steppes of Asia Minor—"up from one place in a whirlwind and down in another place". In the *American Almanac*, 1833–71, it is said that this substance—"unknown to the inhabitants of the region"—was "immediately recognized" by scientists who examined it: and that "the chemical analysis also identified it as a lichen".

This was back in the days when Chemical Analysis was a god. Since then his devotees have been shocked and disillusioned. Just how a chemical analysis could so botanize, I don't know—but it was Chemical Analysis who spoke, and spoke dogmatically. It seems to me that the ignorance of inhabitants, contrasting with the local knowledge of foreign scientists, is overdone: if there's anything good to eat, within any distance conveniently covered by a whirlwind—inhabitants know it. I have data of other falls, in Persia and Asiatic Turkey, of edible substances. They are all dogmatically said to be "manna"; "manna" is dogmatically said to be a species of lichens from the steppes of Asia Minor. The position that I take is that this explanation was evolved in ignorance of the fall of vegetable substances, or edible substances, in other parts of the world: that it is the familiar attempt to explain the general in terms of the local; that, if we shall have data of falls of vegetable substance, in, say, Canada or India, they were not of lichens from the steppes of Asia Minor; that, though all falls in Asiatic Turkey and Persia are sweepingly and conveniently called showers of "manna", they have not been even all of the same substance. In one instance the particles are said to have been "seeds". Though, in *Comptes Rendus*, the substance that fell in 1841 and 1846 is said to have been gelatinous, in the *Bull. Sci. Nat. de Neuchatel*, it is said to have been of something, in lumps the size of a filbert, that had been ground into flour; that of this flour had been made bread, very attractive-looking, but flavourless.

The great difficulty is to explain segregation in these showers—

But deep-sea fishes and occasional falls, down to them, of edible substances; bags of grain, barrels of sugar; things that had not been whirled up from one part of the ocean-bottom, in storms or submarine disturbances, and dropped somewhere else—

I suppose one thinks—but grain in bags never has fallen—

Object of Amherst—its covering like "milled cloth"—

Or barrels of corn lost from a vessel would not sink—but a host of them clashing together, after a wreck—they burst open; the corn sinks, or does when saturated; the barrel staves float longer—

If there be not an overhead traffic in commodities similar to our own commodities carried over this earth's oceans—I'm not the deep-sea fish I think I am.

I have no data other than the mere suggestion of the Amherst object of bags or barrels, but my notion is that bags and barrels from a wreck on one of this earth's oceans, would, by the time they reached the bottom, no longer be recognizable as bags or barrels; that, if we can have data of the fall of fibrous material that may have been cloth or paper or wood, we shall be satisfactory and grotesque enough.

Proc. Roy. Irish Acad., 1–379:

"In the year 1686 some workmen, who had been fetching water from a pond, seven German miles from Memel, on returning to their work after dinner (during which there had been a snowstorm) found the flat ground around the pool covered with a coal-black, leafy mass; and a person who lived near said he had seen it fall like flakes with the snow."

Some of these flake-like formations were as large as a table-top.

"The mass was damp and smelt disagreeably, like rotten seaweed, but when dried, the smell went off."

"It tore fibrously, like paper."

Classic explanation:

"Up from one place, and down in another."

But what went up, from one place, in a whirlwind? Of course, our Intermediatist acceptance is that had this been the strangest substance conceivable, from the strangest other world that could be thought of; somewhere upon this earth there must be a substance similar to it, or from which it would, at least subjectively, or according to description, not be easily distinguishable. Or that everything in New York City is only another degree or aspect of something, or combination of things, in a village of Central Africa. The novel is a challenge to vulgarization: write something that looks new to you: someone will point out that the thrice-accursed Greeks said it long ago. Existence is Appetite: the gnaw of being; the one attempt of all things to assimilate to some higher attempt. It was cosmic that these scientists, who had surrendered to and submitted to the Scientific System, should, consistently with the principles of that system, attempt to assimilate the substance that fell at Memel with some known terrestrial product. At the meeting of the Royal Irish Academy it was brought out that there is a substance, or rather rare occurrence that has been known to form in thin sheets upon marsh land.

It looks like greenish felt.

The substance of Memel:

Damp, coal-black, leafy mass.

But if broken up, the marsh-substance is flakelike, and it tears fibrously.

An elephant can be identified as a sunflower—both have long stems. A camel is indistinguishable from a peanut—if only their humps be considered.

Trouble with this book is that we'll end up a lot of intellectual roués: we'll be incapable of being astonished with anything. We knew to start with, that science and imbecility are continuous; nevertheless so many expressions of the merging-point are at first startling. We think that Prof. Hitchcock's performance in identifying the Amherst phenomenon as a fungus was rather notable as scientific vaudeville, if we acquit him of the charge of seriousness—or that, in a place where fungi were so common that, before a given evening two of them sprang up, only he, a stranger in this fungiferous place, knew a fungus when he saw something like a fungus—if we disregard its quick liquefaction, for instance. It was only a monologue, however: now we have an all-star cast: and they're not only Irish; they're royal Irish.

The royal Irishmen excluded "coal-blackness" and included fibrousness: so then that this substance was "Marsh paper", which "had been raised into the air by storms of wind, and had again fallen".

Second act:

It was said that, according to M. Ehrenburg, "the meteor-paper was found to consist partly of vegetable matter, chiefly of conifervae".

Third act:

Meeting of the royal Irishmen: chairs, tables, Irishmen:

Some flakes of marsh-paper were exhibited.

Their composition was chiefly of conifervae.

This was a double inclusion: or it's the method of agreement that logicians make so much of. So no logician would be satisfied with identifying a peanut as a camel, because both had humps: he demands accessory agreement—that both can live a long time without water, for instance.

Now, it's not so very reasonable at least to the free and easy vaudeville standards that, throughout this book, we are considering, to think that a green substance could be snatched up from one place in a whirlwind, and fall as a black substance somewhere else: but the royal Irishmen excluded something else, and it is a datum that was as accessible to them as it is to me:

That, according to Chladni, this was no little, local deposition that was seen to occur by some indefinite person living near a pond somewhere.

It was a tremendous fall from a vast sky-area.

Likely enough all the marsh paper in the world could not have supplied it.

At the same time, this substance was falling "in great quantities", in Norway and Pomerania. Or see Kirkwood, *Meteoric Astronomy*, p. 66:

"Substance like the charred paper fell in Norway and other parts of northern Europe, Jan. 31, 1686."

Or a whirlwind, with a distribution as wide as that, would not acceptably, I should say, have so specialized in the rare substance called "marsh paper". There'd have been falls of fence rails, roofs of houses, parts of trees. Nothing is said of the occurrence of a tornado in northern Europe, in January, 1686. There is record only of this one substance having fallen in various places.

Time went on, but the conventional determination to exclude data of all falls to this earth, except of substances of this earth, and of ordinary meteoric matter, strengthened. *Annals of Philosophy*, 16–68:

The substance that fell in January, 1686, is described as "a mass of black leaves, having the appearance of burnt paper, but harder, and cohering, and brittle".

"Marsh paper" is not mentioned, and there is nothing said of the "conifervare", which seemed so convincing to the royal Irishmen. Vegetable composition is disregarded, quite as it might be by someone who might find it convenient to identify a crook-necked squash as a big fishhook.

Meteorites are usually covered with a black crust, more or less scale-like. The substance of 1686 is black and scale-like. If so be convenience, "leaf-likeness" is "scale-likeness". In this attempt to assimilate with the conventional, we are told that the substance is a mineral mass: that it is like the black scales that cover meteorites.

The scientist who made this "identification" was Von Grotthus. He had appealed to the god Chemical Analysis. Or the power and glory of mankind—with which we're not always so impressed—but the gods must tell us what we want them to tell us. We see again that, though nothing has identity of its own, anything can be "identified" as anything. Or there's nothing that's not reasonable, if one snoopeth not into its exclusions. But here the conflict did not end. Berzelius examined the substance. He could not find nickel in it. At that time, the presence of nickel was the "positive" standard of judgment against him. Von Grotthus revoked his "identification". (*Annals and Mag. of Nat. Hist.*, 1–3–185.)

This equalization of eminences permits us to project with our own expression, which, otherwise, would be subdued into invisibility:

That it's too bad that no one ever looked to see—hieroglyphics?—something written upon these sheets of paper?

If we have no very real variety of substances that have fallen to this earth; if, upon this earth's surface there is infinite variety of substances detachable by whirlwinds, two falls of such a rare substance as marsh paper would be remarkable.

A writer in the *Edinburgh Review*, 87–194, says that, at the time of writing, he had before him a portion of a sheet of 200 square feet, of a substance that had fallen at Carolath, Silesia, in 1839—exactly similar to cotton-felt, of which clothing might have been made. The god Microscopic Examination had spoken. These substances consisted chiefly of conifervae.

Jour. Asiatic Soc. of Bengal, 1847–pt. 1–193:

That March 16, 1846—about the time of a fall of edible substance in Asia Minor—an olive-grey powder fell at Shanghai. Under the microscope, it was seen to be an aggregation of hairs of two kinds, black ones and rather thick white ones. They were supposed to be mineral fibres, but, when burned, they gave out "the common ammoniacal smell and smoke of burnt hair or feathers". The writer described the phenomenon as "a cloud of 3,800 square miles of fibres, alkali, and sand". In a postscript, he says that other investigators with more powerful microscopes, gave opinion that the fibres were not hairs; that the substance consisted chiefly of conifervae.

Or the pathos of it, perhaps; or the dull and uninspired, but courageous persistence of the scientific: everything seemingly found out is doomed to be subverted—by more powerful microscopes and telescopes; by more refined, precise, searching means and methods—the new pronouncements irrepressibly bobbing up; their reception always as Truth at last; always the illusion of the final; very little of the Intermediatist spirit—

That the new that has displaced the old will itself some day be displaced; that it, too, will be recognized as myth-stuff—

But that if phantoms climb, spooks of ladders are good enough for them.

Annual Register, 1821–681:

That, according to a report by M. Lainé, French Consul at Pernambuco, early in October, 1821, there was a shower of a substance resembling silk. The quantity was as tremendous as might be a whole cargo, lost somewhere between Jupiter and Mars, having drifted around perhaps for centuries, the original fabrics slowly disintegrating. In *Annales de Chimie*, 2–15–427, it is said that samples of this substance were sent to France by M. Lainé, and that they proved to have some resemblances to silky filaments which, at certain times of the year, are carried by the wind near Paris.

In the *Annals of Philosophy*, n.s., 12–93, there is mention of a fibrous substance like blue silk that fell near Naumberg, March 23, 1665. According to Chladni (*Annales de Chimie*, 2–31–264), the quantity was great. He places a question mark before the date.

One of the advantages of Intermediatism is that, in the oneness of quasiness, there can be no mixed metaphors. Whatever is acceptable of anything, is, in some degree or aspect, acceptable of everything. So it is quite proper to speak, for instance, of something that is as firm as a rock and that sails in a majestic march. The Irish are good monists: they have of course been laughed at for their keener perceptions. So it's a book we're writing, or it's a procession, or it's a museum, with the chamber of Horrors rather over-emphasized. A rather horrible correlation occurs in the *Scientific American*, 1859–178. What interests us is that a correspondent saw a silky substance fall from the sky—there was an aurora borealis at the time—he attributes the substance to the aurora.

Since the time of Darwin, the classic explanation has been that all silky substances that fall from the sky are spider webs. In 1832, aboard the *Beagle*, at the mouth of La Plata River, 60 miles from land, Darwin saw an enormous number of spiders, of the kind usually known as "gossamer" spiders, little aeronauts that cast out filaments by which the wind carries them.

It's difficult to express that silky substances that have fallen to this earth were not spider webs. My own acceptance is that spider webs are the merger; that there have been falls of an externally derived silky substance, and also of the webs, or strands, rather, of aeronautic spiders indigenous to this earth: that in some instances it is impossible to distinguish one from the other. Of course, our expression upon silky substances will merge away into expressions upon other seeming textile substances, and I don't know how much better off we'll be—

Except that, if fabricable materials have fallen from the sky—

Simply to establish acceptance of that may be doing well enough in this book of first and tentative explorations.

In *All the Year Round*, 8–254, is described a fall that took place in England, Sept. 21, 1741, in the towns of Bradly, Selborne, and Alresford, and in a triangular space included by these three towns, the substance is described as "cobwebs"—but it fell in flake-formation, or in "flakes or rags about one inch broad and five or six inches long". Also these flakes were of a relatively heavy substance—"they fell with some velocity". The quantity was great—the shortest side of the triangular space is eight miles long. In the *Wernerian Nat. Hist. Soc. Trans.*, 5–386, it is said that there were two falls—that they were some hours apart—a datum that is becomimg familiar to us—a datum that cannot be taken into the fold, unless we find it repeated over and over again. It is said that the second fall lasted from nine o'clock in the morning until night.

Now the hypnosis of the classic—that what we call intelligence is only an expression of inequilibrium; that when mental adjustments are made, intelligence ceases—or, of course, that intelligence is the confession of ignorance. If you have intelligence upon any subject, that is something you're still learning—if we agree that that which is learned is always mechanically done—in quasi-terms, of course, because nothing is ever finally learned.

It was decided that this substance was spiders' web. That was adjustment. But it's not adjustment to me; so I'm afraid I shall have some intelligence in this matter. If I ever arrive at adjustment upon this subject, then, upon this subject, I shall be able to have no thoughts, except routine-thoughts. I haven't yet quite decided absolutely everything, so I am able to point out:

That this substance was of quantity so enormous that it attracted wide attention when it came down—

That it would have been equally noteworthy when it went up—

That there is no record of anyone, in England or elsewhere, having seen tons of "spider webs" going up, September, 1741.

Further confession of intelligence upon my part:

That, if it be contested, then, that the place of origin may have been far away, but still terrestrial—

Then it's that other familiar matter of incredible "Marksmanship" again—hitting a small, triangular space for hours—interval of hours—then from nine in the morning until night: the same small triangular space.

These are the disregards of the classic explanation. There is no mention of spiders having been seen to fall, but a good inclusion is that, though this substance fell in good-sized flakes of considerable weight, it was viscous. In this respect it was like cobwebs: dogs nosing it on the grass, were blind-folded with it. This circumstance does strongly suggest cobwebs—

Unless we can accept that, in regions aloft there are vast viscous or gelatinous areas; and that things passing through become daubed. Or perhaps we clear up the confusion in the descriptions of the substance that fell in 1841 and 1846, in Asia Minor, described in one publication as gelatinous, and in another as a cereal—that it was a cereal that had passed through a gelatinous region. That the paper-like substance of Memel may have had such an experience may be indicated in that Ehrenberg found in it gelatinous matter, which he called "nostoc". (*Annals and Mag. of Nat. Hist.*, 1–3–185.)

Scientific American, 45–337:

Fall of a substance described as "cobwebs," latter part of October, 1881, in Milwaukee, Wis., and other towns: other towns mentioned are Green Bay, Vesburge, Fort Howard, Sheboygan, and Ozaukee. The aeronautic spiders are known as "gossamer" spiders, because of the extreme lightness of the filaments that they cast out to the wind. Of the substance that fell in Wisconsin, it is said:

"In all instances the webs were strong in texture and very white."

The Editor says:

"Curiously enough, there is no mention in any of the reports that we have seen, of the presence of spiders."

So our attempt to divorce a possible external produce from its terrestrial merger: then our joy of the prospector who thinks he's found something:

The *Monthly Weather Review*, 26–566, quotes the *Montgomery* (Ala.) *Advertiser*:

That, upon Nov. 21, 1898, numerous batches of spiderweb-like substance fell in Montgomery, in strands and in occasional masses several inches long and several inches broad. According to the writer, it was not spiders' web, but something like asbestos; also that it was phosphorescent.

The Editor of the *Review* says that he sees no reason for doubting that these masses were cobwebs.

La Nature, 1883–342:

A correspondent writes that he sends a sample of a substance said to have fallen at Montussan (Gironde), Oct. 16, 1883. According to a witness, quoted by the correspondent, a thick cloud, accompanied by rain and a violent wind, had appeared. This cloud was composed of a woolly substance in lumps the size of a fist, which fell to the ground. The Editor (Tissandier) says of this substance that it was white, but was something that had been burned. It was fibrous. M. Tissandier astonishes us by saying that he cannot identify this substance. We thought that anything could be "identified" as anything. He can say only that the cloud in question must have been an extraordinary conglomeration.

Annual Register, 1832–447:

That, March, 1832, there fell, in the fields of Kourianof, Russia, a combustible yellowish substance, covering, at least two inches thick an area of 600 or 700 square feet. It was resinous and yellowish: so one inclines to the conventional explanation that it was pollen from pine trees—but, when torn, it had the tenacity of cotton. When placed in water, it had the consistency of resin. "This resin had the colour of amber, was elastic like India rubber and smelled like prepared oil mixed with wax."

So in general our notion of cargoes of food supplies:

In *Philosophical Transactions*, 19–224, is an extract from a letter by Mr. Robert Vans, of Kilkenny, Ireland, dated Nov. 15, 1695: that there had been "of late", in the counties of Limerick and Tipperary, showers of a sort of matter like butter or grease ... having "a very stinking smell".

There follows an extract from a letter by the Bishop of Cloyne, upon "a very odd phenomenon", which was observed in Munster and Leinster: that for a good part of the spring of 1695 there fell a substance which the country people called "butter"—"soft, clammy, and a dark yellow"—that cattle fed "indifferently" in fields where this substance lay.

"It fell in lumps as big as the end of one's finger." It had a "strong ill scent". His Grace calls it a "stinking dew".

In Mr. Vans' letter, it is said that the "butter" was supposed to have medicinal properties, and "was gathered in pots and other vessels by some of the inhabitants of this place".

And:

In all the following volumes of *Philosophical Transactions* there is no speculation upon this extraordinary subject. Ostracism. The fate of this datum is a good instance of damnation, not by denial, and not explaining away, but by simple disregard. The fall is listed by Chladni, and is mentioned in other catalogues, but, from the absence of all inquiry, and of all but formal mention, we see that it has been under excommunication as much as was ever anything by the preceding system. The datum has been buried alive. It is as irreconcilable with the modern system of dogmas as ever were geologic strata and vermiform appendix with the preceding system—

If, intermittently, or "for a good part of the spring", this substance fell in two Irish provinces, and nowhere else, we have, stronger than before, a sense of a stationary region overhead, or a region that receives products like this earth's gravitational and meteorological forces are relatively inert – if for many weeks a good part of this substance did hover

75

before finally falling. We suppose that, in 1685, Mr. Vans and the Bishop of Cloyne could describe what they saw as well as could witnesses in 1885: nevertheless, it is going far back; we shall have to have many modern instances before we can accept.

As to other falls, or another fall, it is said in the *Amer. Jour. Sci.*, 1–28–361, that, April 11, 1832—about a month after the fall of the substance of Kourianof—fell a substance that was wine-yellow, transparent, soft, and smelling like rancid oil. M. Herman, chemist who examined it, named it "sky oil". For analysis and chemic reactions, see the *Journal*. *The Edinburgh New Philosophical Journal* 13–368 mentions as "unctuous" substance that fell near Rotterdam, in 1832. In *Comptes Rendus*, 13–215, there is an account of an oily, reddish matter that fell at Genoa, February, 1841.

Whatever it may have been—

Altogether, most of our difficulties are problems that we should leave to later developers of super-geography, I think. A discoverer of America should leave Long Island to someone else. If there be, plying back and forth from Jupiter and Mars and Venus, super-constructions that are sometimes wrecked we think of fuel as well as cargoes. Of course the most convincing data would be of coal falling from the sky: nevertheless, one does suspect that oil-burning engines were discovered ages ago in more advanced worlds—but, as I say, we should leave something to our disciples—so we'll not especially wonder whether these butter-like or oily substances were food or fuel. So we merely note that in the *Scientific American*, 24–323, is an account of hail that fell, in the middle of April, 1871, in Mississippi, in which was a substance described as turpentine.

Something that tasted like orange water, in hailstones about the first of June, near Nîmes, France; identified as nitric acid (*Jour. de Pharmacie*, 1845–272).

Hail and ashes, in Ireland, 1755 (*Sci. Amer.*, 5–168).

That, at Elizabeth, N.J., June 9, 1874, fell hail in which was a substance said by Prof. Leeds of Stevens Institute, to be carbonate of soda (*Sci. Amer.*, 30–262).

We are getting a little away from the lines of our composition, but it will be an important point later that so many extraordinary falls have occurred with hail. Or—if there were of substances that had had origin upon some other part of this earth's surface—had the hail, too, that origin? Our acceptance here will depend upon the number of instances. Reasonably enough, some of the things that fall to this earth should coincide with falls of hail.

As to vegetable substances in quantities so great as to suggest lost cargoes, we have a note in the *Intellectual Observer*, 3–468: That, upon the first of May, 1863, a rain fell at Perpignan, "bringing down with it a red substance which proved on examination to be a red meal mixed with fine sand". At various points along the Mediterranean, this substance fell.

There is, in *Philosophical Transactions*, 16–281, an account of a seeming cereal, said to have fallen in Wiltshire, in 1686—said that some of the "wheat" fell "enclosed in hailstones"—but the writer in *Transactions*, says that he had examined the grains and that they were nothing but seeds of ivy berries dislodged from holes and chinks where birds had hidden them. If birds still hide ivy seeds, and if winds still blow, I don't see why the phenomenon has not repeated in more than two hundred years since.

Or the red matter in rain, at Siena, Italy, May, 1830; said, by Arago to have been vegetable matter (Arago, *Oeuvres*, 12–468).

Somebody should collect data of falls at Siena alone.

In the *Monthly Weather Review*, 29–465, a correspondent writes that, upon Feb. 16, 1901, at Pawpaw, Michigan, upon a day that was so calm that his windmill did not turn, fell a brown dust that looked like vegetable matter. The Editor of the *Review* concludes that this was no widespread fall from a tornado, because it had been reported from nowhere else.

Rancidness—putridity—decomposition—a note that has been struck many times. In a positive sense, of course, nothing means anything, or every meaning is continuous with all other meanings: or that all evidences of guilt, for instance, are just as good evidences of innocence—but this condition seems to mean—things lying around among the stars a long time. Horrible disaster in the time of Julius Caesar; remains from it not reaching this earth till the time of the Bishop of Cloyne: we leave to later research the discussion of bacterial action and decomposition, and whether bacteria could survive in what we call space, of which we know nothing—

Chemical News, 35–183:

Dr. A. T. Machattie, F.C.S., writes that, at London, Ontario, Feb. 24, 1868, in a violent storm, fell, with snow a dark-coloured substance estimated at 500 tons, over a belt 50 miles by 10 miles. It was examined under a microscope by Dr. Machattie who found it to consist mainly of vegetable matter "far advanced in decomposition". The substance was examined by Dr. James Adams, of Glasgow, who gave his opinion that it was the remains of cereals. Dr. Machattie points out that for months before this fall the ground of Canada had been frozen, so that in this case a more than ordinarily remote origin has to be thought of. Dr. Machattie thinks the origin to the south. "However," he says, "this is mere conjecture."

Amer. Jour. Sci., 1841–40:

That, March 24, 1840—during a thunderstorm—at Rajkit, India, occurred a fall of grain. It was reported by Col. Sykes, of the British Association.

The natives were greatly excited—because it was grain of a kind unknown to them.

Usually comes forward a scientist who knows more of the things that natives know best than the natives know—but it so happens that the usual thing was not done definitely in this instance:

"The grain was shown to some botanists, who did not immediately recognize it, but thought it to be either a spartium or a vicia."

Chapter 6

Lead, silver, diamonds, glass.

They sound like the accursed, but they're not: they're now of the chosen—that is, when they occur in metallic or stony masses that Science has recognized as meteorites. We find that resistance is to substances not so mixed in or incorporated.

Of accursed data, it seems to me that punk is pretty damnable. In the *Report of the British Association* 1878–376, there is mention of a light chocolate-brown substance that has fallen with meteorites. No particulars given; not another mention anywhere else that I can find. In this English publication, the word "punk" is not used; the substance is called "amadou". I suppose, if the datum has anywhere admitted to French publication, the word "amadou" has been avoided, and "punk" used.

Or oneness of allness: scientific works and social registers: a Goldstein who can't get in as Goldstein, gets in as Jackson.

The fall of sulphur from the sky has been especially repulsive to the modern orthodoxy—largely because of its associations with the superstitions or principles of the preceding orthodoxy—stories of devils: sulphurous exhalations. Several writers have said that they have had this feeling. So the scientific reactionists, who have rabidly fought the preceding, because it was the preceding: and the scientific prudes, who, in sheer exclusionism, have held lean hands over pale eyes, denying falls of sulphur. I have many notes upon the sulphorous odour of meteorites, and many notes upon phosphorescence of things that come from externality. Some day I shall look over old stories of demons that have appeared sulphurously upon this earth, with the idea of expressing that we have often had undesirable visitors from other worlds; or that an indication of external derivation is sulphurousness. I expect some day to rationalize demonology, but just at present we are scarcely far enough advanced to go so far back.

For a circumstantial account of a mass of burning sulphur, about the size of a man's fist, that fell at Pultusk,

Poland, Jan. 30, 1868, upon a road, where it was stamped out by a crowd of villagers see *Rept. Brit. Assoc.*, 1874–272.

The power of the exclusionists lies in that in their stand are combined both modern and archaic systematists. Falls of sandstone and limestone are repulsive to both theologians and scientists. Sandstone and limestone suggest other worlds upon which occur processes like geological processes; but limestone, as a fossiliferous substance, is of course especially of the unchosen.

In *Science*, March 9, 1888, we read of a block of limestone, said to have fallen near Middleburg, Florida. It was exhibited at the Sub-tropical Exposition, at Jacksonville. The Writer, in *Science*, denies that it fell from the sky. His reasoning is:

There is no limestone in the sky;

Therefore this limestone did not fall from the sky.

Better reasoning I cannot conceive—because we see that a final major premise—universal—true—would include all things: that, then, would leave nothing to reason about—so then that all reasoning must be based upon "something" not universal, or only a phantom intermediate to the two finalities of nothingness and allness, or negativeness and positiveness.

La Nature, 1890–2–127:

Fall, at Pel-et-Der (L'Aube), France, June 6, 1890, of limestone pebbles. Identified with limestone at Château-Landor—or up and down in a whirlwind. But they fell with hail—which, in June, could not very well be identified with ice from Château-Landon. Coincidence, perhaps.

Upon page 70, *Science Gossip*, 1887, the Editor says, of a stone that was reported to have fallen at Little Lever, England, that a sample had been sent to him. It was sandstone. Therefore it had not fallen, but had been on the ground in the first place. But upon page 140, *Science Gossip*, 1887, is an account of "a large, smooth water-worn gritty sandstone pebble" that had been found in the wood of a full-grown beech tree. Looks to me as if it had fallen red-hot and had penetrated the tree with high velocity. But I have never heard of anything falling red-hot from a whirlwind—

The wood around this sandstone pebble was black, as if charred.

Dr. Farrington, for instance, in his books, does not even

mention sandstone. However, the British Association, though reluctant, is less exclusive: Report of 1860, p. 197: substance about the size of a duck's egg, that fell at Raphoe, Ireland, June 9, 1860—date questioned. It is not definitely said that this substance was sandstone, but that it "resembled" friable sandstone.

Falls of salt have occurred often. They have been avoided by scientific writers, because of the dictum that only water and not substances held in solution, can be raised by evaporation. However, falls of salty water have received attention from Dalton and others and have been attributed to whirlwinds from the sea. This is so reasonably contested quasi-reasonably—as to places not far from the sea—

But the fall of salt that occurred high in the mountains of Switzerland—

We could have predicted that the datum could be found somewhere. Let anything be explained in local terms of the coast of England—but also has it occurred high in the mountains of Switzerland.

Large crystals of salt fell—in a hailstorm—Aug. 20, 1870, in Switzerland. The orthodox explanation is a crime: whoever made it, should have had his finger-prints taken. We are told (*An Rec. Sci.*, 1872) that these objects of salt "came over the Mediterranean from some part of Africa".

Or the hypnosis of the conventional—provided it be glib. One reads such an assertion, and provided it be suave and brief and conventional, one seldon questions—or thinks "very strange" and then forgets. One has an impression from geography lessons: Mediterranean not more than three inches wide, on the map; Switzerland only a few more inches away. These sizeable masses of salt are described in the *Amer. Jour. Sci.*, 3-3-239, "essentially imperfect cubic crystals of common salt". As to occurrence with hail—that can in one, or ten, or twenty, instances be called a coincidence.

Another datum: extraordinary year 1883:

London *Times*, Dec. 25, 1883:

Translation from a Turkish newspaper; a substance that fell at Scutari, Dec. 2, 1883, described as an unknown substance, in particles—or flakes?—like snow. "It was found to be saltish to the taste, and to dissolve readily in water."

Miscellaneous:

"Black capillary matter" that fell, Nov. 16, 1857, at Charleston, S.C. (*Amer. Jour. Sci.*, 2–31–459).

Fall of small, friable, vesicular masses, from size of a pea to size of a walnut, at Lobau, Jan. 18, 1835 (*Rept. Brit. Assoc.*, 1860–85).

Objects that fell at Peshawur, India, June, 1893, during a storm: substance that looked like crystallized nitre, and that tasted like sugar (*Nature*, July 13, 1893).

I suppose sometimes deep-sea fishes have their noses bumped by cinders. If their regions be subjacent to Cunard or White Star routes, they're especially likely to be bumped. I conceive of no inquiry: they're deep-sea fishes.

Or the slag of Slains. That it was a furnace-product. The Rev. James Rust seemed to feel bumped. He tried in vain to arouse inquiry.

As to a report, from Chicago, April 9, 1879, that slag had fallen from the sky, Prof. E. S. Bastian (*Amer. Jour. Sci.*, 3–18–78) says that the slag "had been on the ground in the first place". It was furnace-slag. "A chemical examination of the specimens has shown that they possess none of the characteristics of true meteorites."

Over and over and over again, the universal delusion; hope and despair of attempted positivism; that there can be real criteria, or distinct characteristics of anything. If anybody can define—not merely suppose, like Prof. Bastian, that he can define—the true characteristics of anything, or so localize trueness anywhere, he makes the discovery for which the cosmos is labouring. He will be instantly translated, like Elijah, into the Positive Absolute. My own notion is that, in a moment of super-concentration, Elijah became so nearly a real prophet that he was translated to heaven, or to the Positive Absolute, with such velocity that he left an incandescent train behind him. As we go along, we shall find the "true test of meteoritic material," which in the past has been taken as an absolute, dissolving into almost utmost nebulosity. Prof. Bastian explains mechanically, or in terms of the usual reflexes to all reports of unwelcome substances: that near where the slag had been found, telegraph wires had been struck by lighning; that particles of melted wire had been seen to fall near the slag—which had been on the ground in the first place. But, according to the *New York Times*, April 14, 1879, about two bushels of this substance had fallen.

Something that was said to have fallen at Darmstadt, June 7, 1846; listed by Greg *(Rept. Brit. Assoc.,* 1867–416) as "only slag".

Philosophical Magazine, 4–10–381:

That, in 1855, a large stone was found far in the interior of a tree, in Battersea Fields.

Sometimes cannon balls are found embedded in trees. Doesn't seem to be anything to discuss; doesn't seem discussable that any one would cut a hole in a tree and hide a cannon ball, which one could take to bed, and hide under one's pillow, just as easily. So with the stone of Battersea Fields. What is there to say, except that it fell with high velocity and embedded in the tree? Nevertheless, there was a great deal of discussion—

Because, at the foot of the tree, as if broken off the stone, fragments of slag were found.

I have nine other instances.

Slag and cinders and ashes, and you won't believe, and neither will I, that they came from the furnaces of vast aerial super-structions. We'll see what looks acceptable.

As to ashes, the difficulties are great, because we'd expect many falls of terrestrial derived ashes—volcanoes and forest fires.

In some of our acceptances, I have felt a little radical—

I suppose that one of our main motives is to show that there is, in quasi-existence, nothing but the preposterousness and final reasonableness—that the new is the obviously preposterous; that it becomes the established and disguisedly preposterous; that it is displaced, after a while, and is again seen to be the preposterous. Or that all progress is from the outrageous to the academic or sanctified, and back to the outrageous—modified, however, by a trend of higher and higher approximation to the impreposterous. Sometimes I feel a little more uninspired than at other times, but I think we're pretty well accustomed now to the oneness of allness; or that the methods of science in maintaining its system are as outrageous as the attempts of the damned to break in. In the *Annual Record of Science,* 1875–241, Prof. Daubrée is quoted: that ashes that had fallen in the Azores had come from the Chicago fire—

Or the damned and the saved, and there's little to choose

between them; and angels are beings that have not obviously barbed tails to them—or never have such bad manners as to stroke an angel below the waist-line.

However this especial outrage was challenged: the Editor of the *Record* returns to it, in the issue of 1876: considers it "in the highest degree improper to say that the ashes of Chicago were landed in the Azores".

Bull. Soc. Astro, de France, 22–245:

Account of a white substance, like ashes, that fell at Annoy, France, March 27, 1908: simply called a curious phenomenon; no attempt to trace to a terrestrial source.

Flake formations, which may signify passage through a region of pressure, are common; but spherical formations—as if of things that have rolled and rolled along planar regions somewhere—are commoner:

Nature, Jan. 10, 1884, quotes a Kimberley newspaper:

That, toward the close of November, 1883, a thick shower of ashy matter fell at Queenstown, South Africa. The matter was in marble-sized balls, which were soft and pulpy, but which, upon drying, crumbled at touch. The shower was confined to one narrow streak of land. It would be only ordinarily preposterous to attribute this substance to Krakatoa—

But, with the fall, loud noises were heard—

But I'll omit many notes upon ashes; if ashes should sift down upon deep-sea fishes, that is not to say that they came from steamships.

Data of falls of cinders have been especially damned by Mr. Symons, the meteorologist, some of whose investigations we'll investigate later—nevertheless—

Notice of a fall, in Victoria, Australia, April 14, 1875 (*Rept. Brit. Assoc.*, 1875–242)—at least we are told, in the reluctant way, that someone "thought" he saw matter fall near him at night, and the next day found something that looked like cinders.

In the *Proc. of the London Roy. Soc.*, 19–122, there is an account of cinders that fell on the deck of a lightship, Jan. 9, 1873. In the *Amer. Jour. Sci.*, 2–24–449, there is a notice that the Editor had received a specimen of cinders said to have fallen—in showery weather—upon a farm, near Ottowa, Ill., Jan. 17, 1857.

But after all, ambiguous things they are, cinders or ashes or slag or clinkers the high priest of the accursed that must speak aloud for us is—coal that has fallen from the sky.

Or coke:

The person who thought he saw something like cinders, also thought he saw something like coke, we are told.

Nature, 36–119:

Something that "looked exactly like coke" that fell—during a thunderstorm—in the Orne, France, April 24, 1887.

Or charcoal:

Dr. Angus Smith, in the *Lit. and Phil. Soc. of Manchester Memoirs*, 2–9–146, says that, about 1827—like a great deal in Lettell's *Principles* and Darwin's *Origin*, this account is from hear-say—something fell from the sky, near Allport, England. It fell luminously, with a loud report, and scattered in a field. A fragment that was seen by Dr. Smith, is described by him as having "the appearance of a piece of common wood charcoal". Nevertheless, the re-assured feeling of the faithful, upon reading this, is burdened with data of differences: the substance was so uncommonly heavy that it seemed as if it had iron in it; also there was "a sprinkling of sulphur". This material is said, by Prof. Baden-Powell, to be "totally unlike that of any other meteorite". Greg, in his catalogue (*Rept. Brit. Assoc.*, 1860–73), calls it "a more than doubtful substance"—but again, against reassurance, that is not doubt of authenticity. Greg says that it is like compact charcoal with particles of sulphur and iron pyrites embedded.

Reassurance rises again:

Prof. Baden-Powell says: "It contains also charcoal, which might perhaps be acquired from matter among which it fell."

This is a common reflex with the exclusionists: that substances not "truly meteorite" did not fall from the sky, but were picked up by "truly meteorite" things, of course only on their surfaces, by impact with this earth.

Rhythm of reassurances and their declines:

According to Dr. Smith, this substance was not merely coated with charcoal; his analysis gives 43.59 per cent carbon.

Our acceptance that coal has fallen from the sky will be via data of resinous substances and bituminous substances, which merge so that they cannot be told apart.

Resinous substances said to have fallen at Kaba,

Hungary. April 15, 1887 (*Rept Brit. Assoc.*, 1860–94).

A resinous substance that fell after a fireball? at Neuhaus, Bohemia, Dec. 17, 1824 (*Rept. Brit. Assoc.*, 1860–70).

Fall, July 28, 1885, at Luchon during a storm of a brownish substance; very friable, carbonaceous matter; when burned it gave out a resinous odour (*Comptes Rendus* 103–837).

Substance that fell, Feb. 17, 18, 19, 1841, at Genoa, Italy, said to have been resinous; said by Arago (*Oeuvres*, 12–469) to have been bituminous matter and sand.

Fall—during a thunderstorm—July, 1681, near Cape Cod, upon the deck of an English vessel, the *Albemarle*, of "burning, bituminous matter" (*Edin. New Phil. Jour.*, 26–86); a fall, at Christiania, Norway, June 13, 1822, of bituminous matter listed by Greg as doubtful; fall of bituminous matter in Germany, March 8, 1798 listed by Greg. Lockyer (*The Meteoric Hypothesis* p. 24) says that the substance that fell at the Cape of Good Hope, Oct. 13, 1838—about five cubic feet of it: substance so soft that it was cuttable with a knife—"after being experimented upon, it left a residue, which gave out a very bituminous smell".

And this inclusion of Lockyer's—so far as findable in all books that I have read—is, in books, about as close as we can get to our desideratum—that coal had fallen from the sky. Dr. Farrington, except with a brief mention, ignores the whole subject of the fall of carbonaceous matter from the sky. Proctor, in all of his books that I have read—is, in books, about as close as we can get to the admission that carbonaceous matter has been found in meteorites "in very minute quantities"—or my own suspicion is that it is possible to damn something else only by losing one's own soul—quasi-soul, of course.

Sci. Amer., 35–120:

That the substance that fell at the Cape of Good Hope "resembled a piece of anthracite coal more than anything else".

It's a mistake I think: the resemblance is to bituminous coal—but it is from the periodicals that we must get our data. To the writers of books upon meteorites it would be as wicked—by which we mean departure from the characters of an established species—quasi-established, of course—

to say that coal has fallen from the sky, as would be, to something in a barnyard, a temptation that it climb a tree and catch a bird. Domestic things in a barnyard: and how wild things from forests outside seem to them. Or the homeopathist—but we shall shovel data of coal.

And, if over and over, we shall learn of masses of soft coal that have fallen upon this earth, if in no instance has it been asserted that the masses did not fall, but were upon the ground in the first place if we have many instances, this time we turn down good and hard the mechanical reflex that these masses were carried from one place to another in whirlwinds because we find it too difficult to accept that whirlwinds could so select, or so specialize in a peculiar substance. Among writers of books, the only one I know of who makes more than brief mention is Sir Robert Ball. He represents a still more antique orthodoxy, or is an exclusionist of the old type, still holding out against even meteorites. He cites several falls of carbonaceous matter, but with disregards that make for reasonableness that earth matter may have been caught up by whirlwinds and flung down somewhere else. If he had given a full list, he would be called upon to explain the special affinity of whirlwinds for a special kind of coal. He does not give a full list. We shall have all that's findable and we shall see that against this disease we're writing, the homeopathist's prescription availeth not. Another exclusionist was Prof. Lawrence Smith. His psychotropism was to respond to all reports of carbonaceous matter falling from the sky, by saying that this damned matter had been deposited upon things of the chosen by impact with this earth. Most of our data antedate him, or were contemporaneous with him, or were as accessible to him as to us. In his attempted positivism it is simply—and beautifully—disregarded that, according to Berthelot, Berzelious, Cloez, Wohler and others these masses are not merely coated with carbonaceous matter, but are carbonaceous throughout, or are permeated throughout. How anyone could so resolutely and dogmatically and beautifully and blindly hold out would puzzle us were it not for our acceptance that only to think is to exclude and include; and to exclude some things that have as much right to come in as have the included—that to have an opinion upon any subject is to be a Lawrence

Smith—because there is no definite subject.

Dr. Walter Flight (*Eclectic Magazine*, 89–71) says, of the substance that fell near Alais, France, March 15, 1806, that it "emits a faint bituminous substance" when heated, according to the observations of Bergelius and a commission appointed by the French Academy. This time we have not the reluctances expressed in such words as "like" and "resembling". We are told that this substance is "an earthy kind of coal".

As to "minute quantities" we are told that the substance that fell at the Cape of Good Hope has in it a little more than a quarter of organic matter which, in alcohol, gives the familiar reaction of yellow, resinous matter. Other instances given by Dr. Flight are:

Carbonaceous matter that fell in 1840, in Tennessee; Cranbourne, Australia, 1861; Montauban, France, May 14, 1864 (twenty masses, some of them as large as a human head, of a substance that "resembled a dull-coloured earthy lignite"); Goalpara, India, about 1867 (about 8 per cent of a hydrocarbon); at Ornans, France, July 11, 1868; substance with "an organic combustible ingredient", at Hessle, Sweden, Jan. 1, 1860.

Knowledge, 4–134:

That, according to M. Daubrée, the substance that had fallen in the Argentine Republic, "resembled certain kinds of lignite and boghead coal". In *Comptes Rendus*, 96–1764, it is said that this mass fell, June 30, 1880 in the province Entre Rios, Argentina: that it is "like" brown coal; that it resembles all the other carbonaceous masses that have fallen from the sky.

Something that fell at Grazac, France, Aug. 10, 1885: when burned, it gave out a bituminous odour (*Comptes Rendus*, 104–1771).

Carbonaceous substance that fell at Rajpunta, India Jan. 22, 1911: very friable: 50 per cent of it soluble in water (*Records Geol. Survey of India*, 44–pt. 1–41).

A combustible carbonaceous substance that fell with sand at Naples, March 14, 1818 (*Amer. Jour. Sci.*, 1–1–309).

Sci. Amer. Sup., 29–11798:

That, June 9, 1889, a very friable substance of a deep greenish black, fell at Mighei, Russia. It contained 5 per cent organic matter, which, when powdered and digested in alcohol, yielded, after evaporation, a bright yellow resin. In this mass was 2 per cent of an unknown mineral.

Cinders and ashes and slag and coke and charcoal and coal.

And the things that sometimes deep-sea fishes are bumped by.

Reluctances and the disguises or covered retreats of such words as "like" and "resemble"—or that conditions of Intermediateness abrupt transitions—but that the spirit animating all Intermediateness is to achieve abrupt transitions—because, if anything could finally break away from its origin and environment that would be a real thing—something not merging away indistinguishably and with the surrounding. So all attempt to be original; all attempt to invent something that is more than mere extension or modification of the preceding, is positivism—or that if one could conceive of a device to catch flies, positively different from, or unrelated to, all other devices—up he'd shoot to heaven, or the Positive Absolute—leaving behind such an incandescent train that in one age it would be said that he had gone aloft in a fiery chariot, and in another age that he had been struck by lightning—

I'm collecting notes upon persons supposed to have been struck by lightning. I think that high approximation to positivism has often been achieved—instantaneous translation—residue of negativeness left behind, looking much like effects of a stroke of lightning. Some day I shall tell the story of the *Marie Celeste*—"properly", as the *Scientific American Supplement* would say—mysterious disappearance of a sea captain, his family, and the crew—

Of positivists, by the route of Abrupt Transition, I think that Manet was notable—but that his approximation was held down by his intense relativity to the public—or that it is quite as impositive to flout and insult and defy as it is to crawl and placate. Of course, Manet began with continuity with Courbet and others, and then, between him and Manet there were mutual influences—but the spirit of abrupt difference is the spirit of positivism, and Manet's stand was against the dictum that all lights and shades must merge away suavely into one another and prepare for one another. So a biologist like De Vries represents positivism, or the breaking of Continuity, by trying to conceive of evolution by mutation—against the dogma of indistinguishable gradations by "minute variations". A Copernicus conceives of helio-centricity. Continuity is against him. He is not permitted to break abruptly with the past. He is permitted to publish his work, but only as "an interesting hypothesis".

Continuity—and that all that we call evolution or progress is attempt to break away from it—

That our whole solar system was at one time attempt by planets to break away from a parental nexus and set up as individualities, and, failing, move in quasi-regular orbits that are expressions of relations with the sun and with one another, all having surrendered, being now quasi-incorporated in a higher approximation to system;

Intermediateness in its mineralogic aspects of positivism—or Iron that strove to break away from Sulphur and Oxygen, and be real, homogenous Iron—failing, inasmuch as elemental iron exists only in text-book chemistry;

Intermediateness in its biological aspect of positivism—or the wild, fantastic, grotesque, monstrous things it conceived of, sometimes in a frenzy of effort to break away abruptly from all preceding types—but failing, in the giraffe-effort, for instance, or only caricaturing an antelope—

All things break one relation only by the establishing of some other relation—

All things cut an umbilical cord only to clutch a breast.

So the fight of the exclusionists to maintain the traditional—or to prevent abrupt transition from the quasi-established—fighting so that here, more than a century after meteorites were included, no other notable inclusion has been made, except that of cosmic dust, data of which Nordenskiold made more nearly real than data in opposition.

So, Proctor, for instance, fought and expressed his feeling of the preposterous, against Sir W. H. Thomson's notions of arrival upon earth of organisms on meteorites—

"I can only regard it as a jest" (*Knowledge*, 1–302).

Or that there is nothing but jest—or something intermediate to jest and tragedy;

That ours is not an existence but an utterance;

That Momus is imagining us for the amusement of the gods, often with such success that some of us seem almost alive—like characters in something a novelist is writing;

which often to considerable degree take their affairs away from the novelist—

That Momus is imagining us and our arts and sciences and religions, and is narrating or picturing us as a satire upon the gods' real existence.

Because—with many of our data of coal that has fallen from the sky as accessible then as they are now, and with the scientific pronouncement that coal is fossil, how, in a real existence, by which we mean a consistent existence, or a state in which there is real intelligence, or a form of thinking that does not indistinguishably merge away with imbecility, could there have been such a row as that which was raised about forty years ago over Dr. Hahn's announcement that he had found fossils in meteorites?

Accessible to anybody at that time:

Philosophical Magazine, 4–17–425:

That the substance that fell at Kaba, Hungary, April 15, 1857, contained organic matter "analogous to fossil waxes".

Or limestone:

Of the block of limestone which was reported to have fallen at Middleburg, Florida, it is said (*Science*, 11–118) that, though something had been seen to fall in "an old cultivated field", the witnesses who ran to it picked up something that "had been upon the ground in the first place". The writer who tells us this, with the usual exclusion-imagination known as stupidity, but unjustly, because there is no real stupidity, thinks he can think of a good-sized stone that had for many years been in a cultivated field, but that had never been seen before–had never interfered with ploughing, for instance. He is earnest and unjarred when he writes that this stone weighs 200 pounds. My own notion, founded upon my own experience in seeing, is that a block of stone weighing 500 pounds might be in one's parlour twenty years, virtually unseen—but not in an old cultivated field, where it interfered with plough-ing—not anywhere—if it interfered.

Dr. Hahn said that he had found fossils in meteorites. There is a description of the corals, sponges, shells, and crinoids, all of them microscopic, which he photographed, in *Popular Science*, 20–83.

Dr. Hahn was a well-known scientist. He was better known after that.

Anybody may theorize upon other worlds and conditions upon them that are similar to our own conditions: if his notions be resented undisguisedly as fiction, or only as an "interesting hypothesis", he'll stir up no prude rages.

But Dr. Hahn said definitely that he had found fossils in specified meteorites: also he published photographs of them. His book is in the New York Public Library. In the reproductions every feature of some of the shells is plainly marked. If they're not shells neither are things under an oyster-counter. The striations are very plain: one sees even the hinges where bivalves are joined.

Prof. Lawrence Smith (*Knowledge*, 1–258):

"Dr. Hahn is a kind of half-insane man, whose imagination has run away with him."

Conservation of Continuity.

Then Dr. Weinland examined Dr. Hahn's specimens. He gave his opinion that they are fossils and that they are not crystals of enstatite, as asserted by Prof. Smith, who had never seen them.

The damnation of denial and the damnation of disregard:

After the publication of Dr. Wienland's findings—silence.

Chapter 7

The living things that have come down to this earth:

Attempts to preserve the system:

That small frogs and toads, for instance, never have fallen from the sky, but were—"on the ground, in the first place"; or that there have been such falls—"up from one place in a whirlwind, and down in another".

Were there some especially froggy place near Europe, as there is an especially sandy place, the scientific explanation would of course be that all frogs falling from the sky in Europe come from that centre of frogeity.

To start with, I'd like to emphasize something that I am permitted to see because I am still primitive or intelligent or in a state of maladjustment:

That there is not one report findable of a fall of tadpoles from the sky.

As to "there in the first place":

See *Leisure Hours*, 3–779, for accounts of small frogs, or toads, said to have been seen to fall from the sky. The writer says that all observers were mistaken: that the frogs or toads must have fallen from trees or other places overhead.

Tremendous number of little toads, one or two months old, that were seen to fall from a great thick cloud that appeared suddenly in a sky that had been cloudless, August, 1804, near Toulouse, France, according to a letter from Prof. Pontus to Arago. (*Comptes Rendus*, 3–54.)

Many instances of frogs that were seen to fall from the sky. (*Notes and Queries*, 8–6–104); accounts of such falls, signed by witnesses. (*Notes and Queries*, 8–6–104);

Scientific American, July 12, 1873:

"A shower of frogs which darkened the air and covered the ground for a long distance is the reported result of a recent rainstorm at Kansas City, Mo."

As to having been there "in the first place":

Little frogs found in London, after a heavy storm, July 30, 1838. (*Notes and Queries*, 8–7–437);

Little toads found in a desert, after a rainfall (*Notes and Queries*, 8–8–493).

To start with I do not deny—positively—the conventional explanation of "up and down". I think that there may have been such occurrences. I omit many notes that I have upon indistinguishables. In the London *Times*, July 4, 1883, there is an account of a shower of twigs and leaves and tiny toads in a storm upon the slopes of the Appenines. These may have been the ejectamenta of a whirlwind. I add, however, that I have notes upon two other falls of tiny toads, in 1883, one in France and one in Tahiti; also of fish in Scotland. But in the phenomenon of the Appenines, the mixture seems to me to be typical of the products of a whirlwind. The other instances seem to me to be typical of—something like migration? Their great numbers and their homogeneity. Over and over in these annals of the damned occurs the datum of segregation. But a whirlwind is thought of as a condition of chaos—quasi-chaos: not final negativeness, of course—

Monthly Weather Review, July, 1881:

"A small pond in the track of the cloud was sucked dry, the water being carried over the adjoining fields together with a large quantity of soft mud, which was scattered over the ground for half a mile around."

It is so easy to say that small frogs that have fallen from the sky had been scooped up by a whirlwind; but here are the circumstances of a scoop; in the exclusionist-imagination there is no regard for mud, débris from the bottom of a pond, floating vegetation, loose things from the shores—but a precise picking out of frogs only. Of all instances I have that attribute the fall of small frogs or toads to whirlwinds, only one definitely identifies or places the whirlwind. Also, as has been said before, a pond going up would be quite as interesting as frogs coming down. Whirlwinds we read of over and over—but where and what whirlwind? It seems that anybody who lost a pond would be heard from. In *Symons' Meteorological Magazine*, 32–106, a fall of small frogs, near Birmingham, England, June 30, 1892 attributed to a specific whirlwind—but not a word as to any special pond that had contributed. And something that strikes my attention here is that these frogs are described as almost white.

I'm afraid there is no escape for us: we shall have to give to civilization upon this earth—some new worlds.

Places with white frogs in them.

Upon several occasions we have had data of unknown things that have fallen from—somewhere. But something not to be overlooked is that if living things have landed alive upon this earth—in spite of all we think we know of the accelerative velocity of falling bodies—and have propagated—why the exotic becomes the indigenous, or from the strangest of places we'd expect the familiar. Or if hosts of living frogs have come here—from somewhere else—every living thing upon this earth may, ancestrally, have come from—somewhere else.

I find that I have another note upon a specific hurricane: *Annals and Mag. of Nat. Hist.* 1–3–185:

After one of the greatest hurricanes in the history of Ireland, some fish were found "as far as 15 yards from the edge of a lake".

Have another: this is a good one for the exclusionists:

Fall of fish in Paris: said that a neighbouring pond had been blown dry. (*Living Age*, 52–186.) Date not given, but I have seen it recorded somewhere else.

The best-known fall of fishes from the sky is that which occurred at Mountain Ash, in the Valley of Aberdare, Glamorganshire, Feb. 11, 1859.

The Editor of the *Zoologist*, 2–677, having published a report of a fall of fishes, writes: "I am continually receiving similar accounts of frogs and fishes." But in all the volumes of the *Zoologist*, I can find only two reports of such falls. There is nothing to conclude other than that hosts of data have been lost because orthodoxy does not look favourably upon such reports. *The Monthly Weather Review* records several falls of fishes in the United States; but accounts of these reported occurrences are not findable in other American publications. Nevertheless, the treatment by the *Zoologist* of the fall reported from Mountain Ash is fair. First appears, in the issue of 1859–6493, a letter from the Rev. John Griffith, Vicar of Aberdare, asserting that the fall had occurred, chiefly upon the property of Mr. Nixon, of Mountain Ash. Upon page 6540, Dr. Gray, of the British Museum, bristling with exclusionism, writes that some of these fishes, which had been sent to him alive, were "very young minnows". He says: "On reading the evidence, it seems to me most probably only a practical joke: that one of Mr. Nixon's employees had thrown a pailful of water upon another, who had thought fish in it

95

had fallen from the sky"—had dipped up a pailful from a brook.

Those fishes—still alive—were exhibited at the Zoological Gardens, Regent's Park. The Editor says that one was a minnow and that the rest were sticklebacks.

He says that Dr. Gray's explanation is no doubt right.

But, upon page 6564, he publishes a letter from another correspondent, who apologizes for opposing "so high an authority as Dr. Gray", but says that he had obtained some of these fishes from persons who lived at a considerable distance apart, or considerably out of range of the playful pail of water.

According to the *Annual Register*, 1859–14, the fishes themselves had fallen by pailfuls.

If these fishes were not upon the ground in the first place, we base our objections to the whirlwind explanation upon two data:

That they fell in no such distribution as one could attribute to the discharge of a whirlwind, but upon a narrow strip of land: about 80 yards long and 12 yards wide—

The other datum is again the suggestion that at first seemed so incredible, but for which support is piling up, a suggestion of a stationary source overhead—

That ten minutes later another fall of fishes occurred upon this same narrow strip of land.

Even arguing that a whirlwind may stand still axially, it discharges tangentially. Wherever the fishes came from it does not seem thinkable that some could have fallen and that others could have whirled even a tenth of a minute, then falling directly after the first to fall. Because of these evil circumstances the best adaptation was to laugh the whole thing off and say that someone had soused someone else with a pailful of water in which a few "very young" minnows had been caught up.

In the London *Times*, March 2, 1859, is a letter from Mr. Aaron Roberts, curate of St. Peter's, Carmathon. In this letter the fishes are said to have been about four inches long, but there is some question of species. I think, myself, that they were minnows and sticklebacks. Some persons, thinking them to be sea fishes, placed them in salt water, according to Mr. Roberts. "The effect is stated to have been almost instantaneous death." "Some were placed in fresh water. These seemed to thrive well." As to narrow distribution, we are told that the fishes fell "in and about the premises of Mr. Nixon". "It was not observed at the time that any fish fell in any other part of the neighbourhood, save in the particular spot mentioned."

In the London *Times*, March 10, 1859, Vicar Griffith writes an account:

The evidence of the fall of fish on this occasion was very conclusive. A specimen of the fish was exhibited and was inches long, and that these did not survive the fall.

Report of the British Association, 1859–158:

"The evidence of the fall of fish on this occasion was very conclusive. A specimen of the fish was exhibited and was found to be the *Gasterosteus leirus*.

Gasterosteus is the stickleback.

Altogether I think we have not a sense of total perdition, when we're damned with the explanation that someone soused someone else with a pailful of water in which were thousands of fishes four or five inches long, some of which covered roofs of houses, and some of which remained ten minutes in the air. By way of contrast we offer our own acceptance:

That the bottom of a super-geographical pond had dropped out.

I have a great many notes upon the fall of fishes, despite the difficulty these records have in getting themselves published, but I pick out the instances that especially relate to our super-geographical acceptances, or to the Principles of Super-Geography: or data of things that have been in the air longer than acceptably could a whirlwind carry them; that have fallen with a distribution narrower than is attributable to a whirlwind; that have fallen for a considerable length of time upon the same narrow area of land.

These three factors indicate, somewhere not far aloft, a region of inertness to this earth's gravitation, of course, however, a region that, by the flux and variation of all things, must at times be susceptible—but, afterward, our heresy will bifurcate—

In amiable accommodation to the crucifixion it'll get, I think—

But so impressed are we with the datum that, though there have been reports of small frogs that have fallen from the sky, not one report upon the fall of tadpoles is findable,

97

that to these circumstances another adjustment must be made.

Apart from our three factors of indication, an extraordinary observation is the fall of living things without injury to them. The devotees of St. Isaac explain that they fall upon thick grass and so survive but; Sir James Emerson Tenant, in his *History of Ceylon*, tells of a fall of fishes upon gravel, by which they were seemingly uninjured. Something else apart from our three main interests is a phenomenon that looks like what one might call an alternating series of falls of fishes, whatever the significance may be.

Meerut, India, July, 1824 (*Living Age*, 52–186); Fifeshire, Scotland, summer of 1824 (*Wernerian Nat. Hist. Soc. Trans.*, 5–575; Moradabad, India, July, 1826 (*Living Age*, 52–186); Rosshire, Scotland, 1828 (*Living Age*, 52–186); Moradabad, India, July 20, 1829 (*Lin. Soc. Trans.*, 16–764); Perthshire, Scotland (*Living Age*, 52–186); Argyllshire, Scotland, 1830, March 9, 1830 (*Recreative Science*, 3–339); Feridpoor, India, Feb. 19, 1830 (*Jour. Asiatic Soc. of Bengal*, 2–650).

A psycho-tropism that arises here—disregarding serial significance—or mechanical unintelligent, repulsive reflex—is that the fishes of India did not fall from the sky; that they were found upon the ground after torrential rains, because streams had overflowed and had than receded.

In the region of Inertness that we think we can conceive of, or a zone that is to this earth's gravitation very much like the neutral zone of a magnet's attraction, we accept that there are bodies of water and also clear spaces—bottoms of ponds dropping out—very interesting ponds, having no earth at bottom—vast drops of water afloat in what is called space—fishes and deluges of water falling—

But also other areas, in which fishes—however they got there: a matter that we'll consider—remain and dry, or even putrefy, than sometimes falling by atmospheric dislodgement.

After a "tremendous deluge of rain, one of the heaviest falls on record" (*All the Year Round*, 8–255) at Rajkote, India, July 25, 1850, "the ground was literally covered with fishes".

The word "found" is agreeable to the repulsions of the conventionalists and their concept of an overflowing stream—but, according to Dr. Buist, some of these fishes were "found" on the tops of haystacks.

Ferrel (*A Popular Treatise*, p. 414) tells of a fall of living fishes–some of them having been placed in a tank, where they survived—that occurred in India, about 20 miles south of Calcutta, Sept. 20, 1839. A witness of this fall says:

"The most strange thing which ever struck me was that the fish did not fall helter-skelter, or here and there, but they fell in a straight line, not more than a cubit in breadth." See *Living Age*, 52–186.

Amer. Jour. Sci., 1–32–199:

That according to testimony taken before a magistrate, a fall occurred, Feb. 19, 1830, near Feridpoor, India, of many fishes, of various sizes—some whole and fresh and others "mutilated and putrefying". Our reflex to those who would say that, in the climate of India, it would not take long for fishes to putrefy, is—that high in the air, the climate of India is not torrid. Another peculiarity of this fall is that some of the fishes were much larger than others. Or to those who hold out for segregation in a whirlwind, or that objects, say, twice as heavy as others would be separated from the lighter, we point out that some of these fishes were twice as heavy as others.

In the *Journal of the Asiatic Society of Bengal*, 2–650, depositions of witnesses are given:

"Some of the fish were fresh, but others were rotten and without heads."

"Among the number which I got, five were fresh and the rest stinking and headless."

They remind us of His Grace's observation of some pages back.

According to Dr. Buist, some of these fishes weighed one and a half pounds each and others three pounds.

A fall of fishes at Futtepoor, India, May 16, 1833:

"They were all dead and dry." (Dr. Buit, *Living Age*, 52–186.)

India is far away: about 1830 was long ago.

Nature, Sept. 1918-46:

A correspondent writes, from the Dove Marine Laboratory, Cuttercoats, England, that, at Hindon, a suburb of Sunderland, Aug. 24, 1918, hundreds of small fishes, identified as sand eels, had fallen—

Again the small area: about 60 by 30 yards.

The fall occurred during a heavy rain that was accompanied by thunder—or the indications of disturbance aloft—but by no visible lightning. The sea is close to Hindon, but if you try to think of these fishes having described a trajectory in a whirlwind from the ocean, consider this remarkable datum:

That, according to witnesses, the fall upon this small area occupied ten minutes.

I cannot think of a clearer indication of a direct fall from a stationary source.

And:

"The fish were all dead, and indeed stiff and hard, when picked up, immediately after the occurrence."

By all of which I mean that we have only begun to pile up our data of things that fall from a stationary source overhead: we'll have to take up the subject as rigorously arrived at as ever has been a belief, can emerge from the accursed.

I don't know how much the horse and the barn will help us to emerge; but, if ever anything did go up from this earth's surface and stay up—those damned things may have:

Monthly Weather Review, May 1878:

In a tornado, in Wisconsin, May 23, 1878, "a barn and a horse were carried completely away, and neither horse nor barn, nor any portion of either have since been found".

After that, which would be a little strong were it not for a steady improvement in our digestions that I note as we go along, there is little of the bizarre or the unassimilable in the turtle that hovered six months or so over a small town in Mississippi:

Monthly Weather Review, May, 1894:

That, May 11, 1894, at Vicksburg, Miss., fell a small piece of alabaster; that, at Bovina, eight miles from Vicksburg, fell a gopher turtle. They fell in a hailstorm.

This item was widely copied at the time: for instance, *Nature*, one of the volumes of 1894, page 430, and *Jour. Roy. Met. Soc.*, 20–273. As to discussion—not a word. Or Science and its continuity with Presbyterianism—data like this are damned at birth. The *Weather Review* does sprinkle or baptize, or attempt to save, this infant—but in all the meteorological literature that I have gone through, after that date—not a word, except mention once or twice. The Editor of the *Review* says:

"An examination of the weather map shows that these hailstorms occur on the south side of the region of cold northerly winds, and were but a small part of a series of similar storms; apparently some special local whirls or gusts carried heavy objects from this earth's surface up to the cloud regions."

Of all incredibilities that we have to choose from, I give first place to a notion of a whirlwind pounding upon a region and scrupulously selecting a turtle and a piece of alabaster. This time, the other mechanical thing "there in the first place" cannot rise in response to its stimulus: it is resisted in that these objects were coated with ice—month of May in the southern state. If a whirlwind at all, there must have been very limited selection: there is no record of the fall of other objects. But there is no attempt in the *Review* to specify a whirlwind.

These strangely associated things were remarkably separated.

They fell eight miles apart.

Then—as if there were real reasoning—they must have been high to fall with such divergence, or one of them must have been carried partly horizontally eight miles farther than the other. But either supposition argues for power more than that of a local whirl or gust, or argues for a great, specific disturbance, of which there is no record—for the month of May, 1894.

Nevertheless—as if I really were reasonable—I do feel that I have to accept that this turtle had been raised from this earth's surface, somewhere near Vicksburg—because the gopher turtle is common in the southern states.

Then I think of a hurricane that occurred in the state of Mississippi weeks or months before May 11, 1894.

No—I don't look for it—and inevitably find it.

Or that things can go up so high in hurricanes that they stay up indefinitely—but may, after a while, be shaken down by storms. Over and over have we noted the occurrence of strange falls in storms. So then that the turtle and the piece of alabaster may have had far different origins—from different worlds, perhaps—have entered a region of suspension over this earth—wafting near each other—long duration—final precipitation by atmospheric disturbance—with hail—or that hailstones, too when large, are phenomena of suspension of long duration: that it

101

is highly unacceptable that the very large ones become so great only in falling from the clouds.

Over and over has the note of disagreeableness, or of putrefaction, been struck—long duration. Other indications of long duration.

I think of a region somewhere above this earth's surface in which gravitation is inoperative and is not governed by the square of the distance—quite as magnetism is negligible at a very short distance from a magnet. Theoretically the attraction of a magnet should decrease with the square of the distance, but the falling-off is found to be almost abrupt at a short distance.

I think that things raised from the earth's surface to that region have been held there until shaken down by storms—The Super-Sargasso Sea.

Derelicts, rubbish, old cargoes from inter-planetary wrecks; things cast out into what is called space by convulsions of other planets, things from the times of the Alexanders, Caesars and Napoleons of Mars and Jupiter and Neptune; things raised by this earth's cyclones: horses and barns and elephants and flies and dodoes, moas, and pterodactyls; all, however, tending to disintegrate into homogeneous-looking muds or dusts, red or black or yellow—treasure-troves for the palaeontologists and for the archaeologists—accumulations of centuries—cyclones of Egypt, Greece, and Assyria—fishes dried hard, there a short time: others there long enough to putrefy—

But the omnipresence of Heterogeneity—or living fishes, also—ponds of fresh water: oceans of salt water.

As to the Law of Gravitation, I prefer to take one simple stand:

Orthodoxy accepts the correlation and equivalence of forces:

Gravitation is one of these forces.

All other forces have phenomena of repulsion and of inertness irrespective of distance, as well as of attraction.

But Newtonian Gravitation admits attraction only:

Then Newtonian Gravitation can only be one-third acceptable even to the orthodox, or there is denial of the correlation and equivalence of forces.

Or still simpler:

Here are the data.

Make what you will, yourself, of them.

102

In our Intermediatist revolt against homogeneous, or positive, explanations, or our acceptance that the all-sufficing cannot be less than universality, besides which, however, there would be nothing to suffice, our expression upon the Super-Sargasso Sea, though it harmonizes with data of fishes that fall as if from a stationary source—and, of course, with other data, too—is inadequate to account for two peculiarities of the falls of frogs:

That never has a fall of tadpoles been reported:

That never has a fall of full-grown frogs been reported—Always frogs a few months old.

It sounds positive, but if there be such reports they are somewhere out of my range of reading.

But tadpoles would be more likely to fall from the sky than would frogs, little or big, if such falls be attributed to whirlwinds; and more likely to fall from the Super-Sargasso Sea if, though very tentatively and provisionally, we accept the Super-Sargasso Sea.

Before we take up an especial expression upon the fall of immature and larval forms of life to this earth, and the necessity than of conceiving factor besides mere stationariness or suspension or stagnation, there are other data that are similar to data of falls of fishes.

Science Gossip, 1886–238:

That small snails, of a land species, had fallen near Redruth, Cornwall, July 8, 1886, "during a heavy thunderstorm": roads and fields strewn with them, so that they were gathered up by the hatful: none seen to fall by the writer of this account: snails said to be "quite different to any previously known in this district".

But, upon page 282, we have better orthodoxy. Another correspondent writes that he had heard of the supposed fall of snails: that he had supposed that all such stories had gone the way of witch stories: that, to his astonishment, he had read an account of this absurd story in a local newspaper of "great and deserved repute".

"I thought I should for once like to trace the origin of one of these fabulous tales."

Our own acceptance is that justice cannot be in an intermediate existence, in which there can be approximation only to justice or to injustice; that to be fair is to have no opinion at all; that to be honest is to be uninterested; that to investigate is to admit prejudice; that nobody has ever really investigated anything, but has always sought positively to prove or to disprove something that was conceived of, or suspected, in advance.

"As I suspected," says this correspondent, "I found that the snails were of a familiar land-species"—that they had been upon the ground "in the first place".

He found that the snails had appeared after the rain: that "astonished rustics had jumped to the conclusion that they had fallen".

He met one person who said that he had seen the snails fall.

"This was his error," says the investigator.

In the *Philosophical Magazine*, 58–310, there is an account of snails said to have fallen at Bristol in a field of three acres, in such quantities that they were shovelled up. It is said that the snails "may be considered as a local species". Upon page 457, another correspondent says that the numbers had been exaggerated, and that in his opinion they had been on the ground in the first place. But that there had been some unusual condition aloft comes out in his observation upon "the curious azure-blue appearance of the sun, at the time".

Nature, 47–278:

That, according to *Das Wetter*, December, 1892, upon Aug. 9, 1892, a yellow cloud appeared over Paderborn, Germany. From this cloud fell a torrential rain, in which were hundreds of mussels. There is no mention of whatever may have been upon the ground in the first place, nor of a whirlwind.

Lizards—said to have fallen on the sidewalks of Montreal, Canada, Dec. 28, 1857. (*Notes and Queries*, 8–6–104.)

In the *Scientific American*, 3–112, a correspondent writes, from South Granville, N.Y., that, during a heavy shower, July 3, 1860, he heard a peculiar sound at his feet, and looking down saw a snake lying as if stunned by a fall. It then came to life. Grey snake, about a foot long.

These data have any meaning or lack of meaning or degree of damnation you please: but, in the matter of the fall that occurred at Memphis, Tennessee, occur some strong significances. Our quasi-reasoning upon this subject applies to all segregations so far considered.

Monthly Weather Review, Jan, 15, 1877:

That, in Memphis, Tenn., Jan. 15, 1877, rather strictly localized, or "in a space of two blocks", and after a violent storm in which the rain "fell in torrents", snakes were found. They were crawling on sidewalks, in yards, and in streets, and in masses—but "none were found on roofs or any other elevation above ground" and "none were seen to fall".

If you prefer to believe that the snakes had always been there, or had been upon the ground in the first place, and that it was only that something occurred to call special attention to them, in the streets of Memphis, Jan. 15, 1877—why, that's sensible: that's the common sense that has been against us from the first.

It is not said whether the snakes were of a known species or not, but that "when first seen, they were of dark brown, almost black". Blacksnakes, I suppose.

If we accept that these snakes did fall, even though not seen to fall by all the persons who were out sight-seeing in a violent storm, and had not been in the streets crawling loose or in thick tangled masses, in the first place;

If we try to accept that these snakes had been raised from some other part of this earth's surface in a whirlwind;

If we try to accept that a whirlwind could segregate them—

We accept the segregation of other objects raised in that whirlwind.

Then, near the place of origin, there would have been a fall of heavier objects that had been snatched up with the snakes—stones, fence rails, limbs of trees. Say that the snakes occupied the next gradation, and would be the next to fall. Still farther would there have been separate falls of lightest objects: leaves, twigs, tufts of grass.

In the *Monthly Weather Review* there is no mention of other falls said to have occurred anywhere in January, 1877.

Again ours is the objection against selectiveness by a whirlwind. Conceivably a whirlwind could scoop out a den of hibernating snakes, with stones and earth and an infinitude of other débris, snatching up dozens of snakes—I don't know how many to a den—hundreds maybe—but, according to the account of this occurrence in the *New York Times*, there were thousands of them; alive; from one foot to eighteen inches in length. The *Scientific American*,

105

36–86, records the fall, and says that there were thousands of them. The usual whirlwind-explanation is given—"but in what locality snakes exist is such abundance is yet a mystery".

This matter of enormousness of numbers suggests to me something of a migratory nature—but that snakes in the United States do not migrate in the month of January, if ever.

As to falls or flutterings of winged insects from the sky, prevailing notions of swarming would seem explanatory enough: nevertheless, in instances of ants, there are some peculiar circumstances.

L'Astronomie, 1889–353:

Fall of fishes, June 13, 1889, in Holland; ants, Aug. 1, 1889, Strasbourg; little toads, Aug. 2, 1889, Savoy.

Fall of ants, Cambridge, England, summer of 1874—"some were wingless". (*Scientific American*, 30–193). Enormous fall of ants, Nancy, France, July 21, 1887—"most of them were wingless", (*Nature*, 36–349.) Fall of enormous, unknown ants—size of wasps—Manitoba, June, 1895. (*Sci. Amer.*, 72–385.)

However, our expression will be:

That wingless, larval forms of life, in numbers so enormous that migration from some place external to this earth is suggested, have fallen from the sky.

That these "migrations"—if such can be our acceptance—have occurred at a time of hibernation and burial far in the ground of larvae in the northern latitudes of this earth; that there is significance in recurrence of these falls in the last of January—or that we have the square of an incredibility in such a notion as that of selection of larvae by whirlwinds, compounded with selection of the last of January.

I accept that there are "snow worms" upon this earth—whatever their origin may have been. In the *Proc. Acad. Nat. Sci. of Philadelphia*, 1899–125, there is a description of yellow worms and black worms that have been found together on glaciers in Alaska. Almost positively were there no other form of insect-life upon these glaciers, and there was no vegetation to support insect-life, except microscopic organisms. Nevertheless the decription of this probably polymorphic species fits the description of larvae said to have fallen in Switzerland, and less definitely

106

fits another description. There is no opposition here, if our data of falls are clear. Frogs of every-day ponds look like frogs said to have fallen from the sky—except the whitish frogs of Birmingham. However, all falls of larvae have not positively occurred in the last of January:

London *Times*, April 14, 1837:

That, in the parish of Bramford Speke, Devonshire, a large number of black worms, about three-quarters of an inch in length, had fallen in a snowstorm.

In Timbs' *Year Book*, 1877–26, it is said that, in the winter of 1876, at Christiania, Norway, worms were found crawling upon the ground. The occurrence is considered a great mystery, because the worms could not have come up from the ground, inasmuch as the ground was frozen at the time, and because they were reported from other places, also, in Norway.

Immense number of black insects in a snowstorm, in 1827, at Pakroff, Russia. (*Scientific American*, 30–193.)

Fall, with snow, at Orenburg, Russia, Dec. 14, 1830, of a multitude of small, black insects, said to have been gnats, but also said to have had flea-like motions (*Amer. Jour. Sci.*, 1–22–375.)

Large number of worms found in a snowstorm, upon the surface of snow about four inches thick, near Sangerfield, N.Y., Nov. 18, 1850 (*Scientific American,* 6–96). The writer thinks that the worms had been brought to the surface of the ground by rain, which had fallen previously.

Scientific American, Feb. 21, 1891:

"A puzzling phenomenon has been noted frequently in some parts of Valley Bend District, Randolph County, Va., this winter. The crust of the snow has been covered two or three times with worms resembling the ordinary cut worms. Where they come from, unless they fall with the snow, is inexplicable." In the *Scientific American,* March 7, 1891, the Editor says that similar worms had been seen upon the snow near Utica, N.Y., and in Oneida and Herkimer Counties; that some of the worms had been sent to the Department of Agriculture at Washington. Again two species, or polymorphism. According to Prof. Riley, it was not polymorphism, "but two distinct species"—which, because of our data, we doubt. One kind was larger than the other: colour differences not distinctly stated. One is called the larvae of the common soldier beetle and the other "seems to be a variety of the bronze cut worm". No attempt to explain the occurrence in snow.

Fall of great numbers of larvae of beetles, near Mortagne, France, May, 1858. The larvae were inanimate as if with cold. (*Annales Sociëtë Entomologique de France*, 1858.)

Trans. Ent. Soc. of London, 1871–183, records "snowing of larvae", in Silesia, 1806; "appearance of many larvae on the snow", in Saxony, 1811; "larvae found alive on the snow", 1828; larvae and snow which "fell together", in the Eifel, Jan. 30, 1847; "fall of insects", Jan. 24, 1849, in Lithuania; occurrence of larvae estimated at 300,000 on the snow in Switzerland, 1856. The compiler says that most of these larvae live underground, or at the roots of trees; that whirlwinds uproot trees, and carry away the larvae—conceiving of them as not held in masses of frozen earth—all as neatly detachable as currants in something. In the *Revue et Magasin de Zoologie*, 1849–72, there is an account of the fall in Lithuania, Jan. 24, 1849—that black larvae had fallen in enormous numbers.

Larvae thought to have been of beetles, but described as "caterpillars", not seen to fall, but found crawling on the snow, after a snowstorm, at Warsaw, Jan. 20, 1850. (*All the Year Round*, 8–253.)

Flammarion (*The Atmosphere*, p. 414) tells of a fall of larvae that occurred Jan. 30, 1869, in a snowstorm, in Upper Savoy. "They could not have been hatched in the neighbourhood, for, during the days preceding, the temperature had been very low"; said to have been a species common in the south of France. In *La Science Pour Tous*, 14–183, it is said that with these larvae there were developed insects.

L'Astronomie, 1890–313:

That, upon the last of January, 1890, there fell, in a great tempest, in Switzerland, incalculable numbers of larvae: some black and some yellow; numbers so great that hosts of birds were attracted.

Altogether we regard this as one of our neatest expressions for external origins and against the whirlwind explanation. If an exclusionist says that, in January, larvae were precisely and painstakingly picked out of the frozen ground, in incalculable numbers, he thinks of a tremendous force—disregarding its refinements: then if origin and precipitation be not far apart, what becomes of an infinitude of other débris, conceiving of no time for segregation?

If he thinks of a long translation—all the way from the south of France to Upper Savoy, he may think then of a very fine sorting over by differences of specific gravity—but in such a fine selection, larvae would be separated from developed insects.

As to differences in specific gravity—the yellow larvae that fell in Switzerland, January, 1890, were three times the size of the black larvae that fell with them. In accounts of this occurrence, there is no denial of the fall.

Or that a whirlwind never brought them together and held them together and precipitated them and only them together—

That they came from Genesistrine.

There's no escape from it. We'll be persecuted for it. Take it or leave it—

Genesistrine.

The notion is that there is somewhere aloft a place of origin of life relatively to this earth. Whether it's the planet Genesistrine, or the moon, or a vast amorphous region superjacent to this earth, or an island in the Super-Sargasso Sea, should perhaps be left to the researches of other super—or extra—geographers. That the first unicellular organisms may have come here from Genesistrine—or that men or anthropomorphic beings may have come here before amoebae: that upon Genesistrine, there may have been an evolution expressible in conventional biological terms, but the evolution in modern Japan—induced by external influences, as a whole, upon this earth, has been a process of population by immigration or by bombardment. Some notes I have upon remains of men and animals encysted, or covered with clay or stone, as if fired here as projectiles, I omit now, because it seems to regard the whole phenomenon as a tropism—as a geotropism—probably atavistic, or vestigial, as it were, or something still continuing long after expiration of necessity; that, once upon a time, all kinds of things came here from Genesistrine, but that now only a few kinds of bugs and things, at long intervals, feel the inspiration.

Not one instance have we of tadpoles that have fallen to this earth. It seems reasonable that a whirlwind could scoop up a pond, frogs and all, and cast down the frogs somewhere else: but, then, more reasonable that a whirlwind could scoop up a pond, tadpoles and all—because tadpoles are more numerous in their season than are the frogs in theirs: but the tadpole-season is earlier in the spring, or in a time that is more tempestuous. Thinking in terms of causation—as if there were real causes—our notion is that, if X is likely to cause Y, but is more likely to cause Z, but does not cause Z, X is not the cause of Y. Upon this quasi-sorites, we base our acceptance that the little frogs that have fallen to earth are not products of whirlwinds; that they came from externality, or from Genesistrine.

I think of Genesistrine in terms of biologic mechanics: not that somewhere there are persons who collect bugs in or about the last of January and frogs in July and August, and bombard this earth, anymore than do persons go through northern regions, catching and collecting birds, every autumn then casting them southward.

But atavistic, or vestigial, geotropism in Genesistrine—or a million larvae start crawling, and a million little frogs start hopping—knowing no more what it's all about than we do when we crawl to work in the morning and hop away at night.

I should say, myself, that Genesistrine is a region in the Super-Sargasso Sea, and that parts of the Super-Sargasso Sea have rhythms of susceptibility to this earth's attraction.

Chapter 8

I accept that, when there are storms, the damnedest of excluded, excommunicated things—things that are leprous to the faithful—are brought down—from the Super-Sargasso Sea—or from what for convenience we call the Super-Sargasso Sea—which by no means has been taken into full acceptance yet.

That things are brought down by storms, just as, from the depths of the sea things are brought up by storms. To be sure it is orthodoxy that storms have little, if any, effect below the waves of the ocean—but—of course—only to have an opinion is to be ignorant of, or to disregard a contradiction, or something else that modified an opinion out of distinguishability.

Symons' Meteorological Magazine, 47–180:

That, along the coast of New Zealand, in regions not subject to submarine volcanic action, deep-sea fishes are often brought up by storms.

Iron and stones that fall from the sky; and atmospheric disturbances:

"There is absolutely no connection between the two phenomena." (*Symons*.)

The orthodox belief is that objects moving at planetary velocity would, upon entering this earth's atmosphere, be virtually unaffected by hurricanes; might as well think of a bullet swerved by someone fanning himself. The only trouble with the orthodox reasoning is the usual trouble—its phantom-dominant—its basing upon a myth—data we've had, and more we'll have, of things in the sky having no independent velocity.

There are so many storms, and so many meteors and meteorites that it would be extraordinary if there were no concurrences. Nevertheless, so many of these concurrences are listed by Prof. Baden-Powell (*Rept. Brit. Assoc.*, 1850–54) that one—notices.

See *Rept. Brit. Assoc.*, 1860—other instances.

The famous fall of stones at Siena, Italy, 1794—"in a violent storm".

See *Greg's Catalogues*—many instances. One that stands out is—"bright ball of fire and light in a hurricane in England, Sept. 2, 1786". The remarkable datum here is that this phenomenon was visible forty minutes. That's about 800 times the duration that the orthodox give to meteors and meteorites.

See the *Annual Register*—many instances.

In *Nature*, Oct. 25, 1877, and the London *Times*, Oct, 1787, something that fell in a gale of Oct. 14, 1877, is described as a "huge ball of green fire". This phenomenon is described by another correspondent, in *Nature*, 17–10, and an account of it by another correspondent was forwarded to *Nature* by W. F. Denning.

There are so many instances that some of us will revolt against the insistence of the faithful that it is only coincidence, and accept that there is connection of the kind called causal. If it is too difficult to think of stones and metallic masses swerved from their courses by storms, if they move at high velocity, we think of low velocity, or of things having no velocity at all, hovering a few miles above this earth, dislodged by storms, and falling luminously.

But the resistance is so great here, and "coincidence" so insisted upon that we'd better have some more instances:

Aerolite in a storm at St. Leonards-on-Sea, England, Sept. 17, 1885—no trace of it found (*Annual Register*, 1885); meteorite in a gale, March 1, 1886, described in the *Monthly Weather Review*, March, 1886; meteorite in a thunderstorm, off coast of Greece, Nov. 19, 1899 (*Nature*, 61–111); fall of a meteorite in a storm, July 7, 1883, near Lachine, Quebec (*Monthly Weather Review*, July, 1883): same phenomenon noted in *Nature*, 28–319; meteorite in a whirlwind, Sweden, Sept. 24, 1883 (*Nature*, 29–15).

London Roy. Soc. Proc., 6–276:

A triangular cloud that appeared in a storm, Dec. 17, 1852; a red nucleus, about half the apparent diameter of the moon, and a long tail; visible 13 minutes; explosion of the nucleus.

Nevertheless, in *Science Gossip*, n.s., 6–65, it is said that, though meteorites have fallen in storms, no connection is supposed to exist between the two phenomena, except by the ignorant peasantry.

But some of us peasants have gone through the *Report of the British Association*, 1852. Upon page 239. Dr. Buist, who had never heard of the Super-Sargasso Sea, says that, although it is difficult to trace connection between the phenomena, three aerolites had fallen in five months, in India, during thunderstorms, in 1851 (may have been 1852). For accounts by witnesses, see page 229 of the *Report*.

Or—we are on our way to account for "thunderstorms".

It seems to me that, very strikingly here, is borne out the general acceptance that ours is only an intermediate existence, in which there is nothing fundamental or nothing final to take as a positive standard to judge by.

Peasants believed in meteorites.

Scientists excluded meteorites.

Peasants believe in "thunderstones".

Scientists exclude "thunderstones".

It is useless to argue that peasants are out in the fields, and that scientists are shut up in laboratories and lecture rooms. We cannot take for a real base that, as to phenomena with which they are more familiar, peasants are more likely to be right than are scientists: a host of biologic and meteorologic fallacies of peasants rises against us.

I should say that our "existence" is like a bridge—except that that comparison is in static terms—but like the Brooklyn Bridge, upon which multitudes of bugs are seeking a fundamental—coming to a girder that seems firm and final—but the girder is built upon supports. A support that seems final. But it is built upon underlying structures. Nothing final can be found in all the bridge, because the bridge itself is not a final thing in itself, but is a relationship between Manhattan and Brooklyn. If our "existence" is a relationship between the Positive Absolute and the Negative Absolute, the quest for finality in it is hopeless: everything in it must be relative, if the "whole" is not a whole, but is, itself, a relation.

In the attitude of Acceptance, our pseudo-base is:

Cells of an embryo are the reptilian era of the embryo;

Some cells feel stimuli to take on new appearances.

If it be of the design of the whole that the next era be mammalian, those cells that turn mammalian will be sustained against resistance, by inertia, of all the rest, and will be relatively right, though not finally right, because they, too, in time will have to give way to characters of other eras of higher development.

113

If we are upon the verge of a new era, in which Exclusionism must be overthrown, it will avail thee not to call us baseborn and frowsy peasants.

In our crude, bucolic way, we now offer an outrage upon common sense that we think will some day be an unquestioned commonplace:

That manufactured objects of stone and iron have fallen from the sky:

That they have brought down from a state of suspension, in a region of inertness to this earth's attraction, by atmospheric disturbances.

The "thunderstone" is usually "a beautifully polished, wedge-shaped piece of green stone", says a writer in the *Cornhill Magazine*, 50–517. It isn't: it's likely to be of almost any kind of stone, but we call attention to the skill with which some of them have been made. Of course this writer says it's all superstition. Otherwise he'd be one of us crude and simple sons of the soil.

Conventional damnation is that stone implements, already on the ground—"on the ground in the first place"—are found near where lightning was seen to strike: that are supposed by astonished rustics, or by intelligence of a low order, to have fallen in or with lightning.

Throughout this book, we class a great deal of science with bad fiction. When is fiction bad, cheap, low? If coincidence is overworked. That's one way of deciding. But with single writers coincidence seldom is overworked: we find the excess in the subject at large. Such a writer as the one of the *Cornhill Magazine* tells us vaguely of beliefs of peasants: there is no massing of instance after instance after instance. Here ours will be the method of massing formation.

Conceivably lightning may strike the ground near where there was a wedge-shaped object in the first place: again and again and again: lightning striking ground near wedge-shaped object in China; lightning striking ground near wedge-shaped object in Central Africa; coincidence in France; coincidence in Java; coincidence in South America—

We grant a great deal but note a tendency to restlessness. Nevertheless this is the psycho-tropism of science to all "thunderstones" said to have fallen luminously.

As to greenstone, it is in the island of Jamaica, where the notion is general that axes of a hard greenstone fall from

114

the sky—"during the rains". (*Jour. Inst. Jamaica*, 2–4.) Some other time we shall inquire into this localization of objects of a specific material. "They are of a stone nowhere else to be found in Jamaica." (*Notes and Queries*, 2–8–24.)

In my own tendency to exclude, or in the attitude of one peasant or savage who thinks he is not to be classed with other peasants or savages, I am not very impressed with what natives think. It would be hard to tell why. If the word of a Lord Kelvin carries no more weight, upon scientific subjects, than the word of a Sitting Bull, unless it be in agreement with conventional opinion—I think it must be because savages have bad table manners. However, my snobbishness, in this respect, loosens up somewhat before very widespread belief by savages and peasants. And the notion of "thunderstones" is as wide as geography itself.

The natives of Burma, China, Japan, according to Blinkenberg (*Thunder Weapons*, p. 100)—not, of course, that Blinkenberg accepts one word of it—think that carved stone objects have fallen from the sky, because they think they have seen such objects fall from the sky. Such objects are called "thunderbolts" in these countries. They are called "thunderstones" in Moravia, Holland, Belgium, France, Cambodia, Sumatra, and Siberia. They're called "storm stones" in Lausitz; "sky arrows" in Slavonia; "thunder axes" in England and Scotland; "lightning stones" in Spain and Portugal; "sky axes" in Greece; "lightning flashes" in Brazil; "thunder teeth" in Amboina.

The belief is as widespread as is belief in ghosts and witches, which only the superstitious deny today.

As to beliefs by North American Indians, Tyler gives a list of references (*Primitive Culture*, 2–237). As to South American Indians—"Certain stone hatchets are said to have fallen from the heavens." (*Jour. Amer. Folk Lore*, 17–203.)

If you, too, revolt against coincidence after coincidence after coincidence, but find our interpretation of "thunderstones" just a little too strong or rich for digestion, we recommend the explanation of one, Tallius, written in 1649:

"The naturalists say they are generated in the sky by fulgurous exhalation conglobed in a cloud by the circumfused humour."

Of course the paper in the *Cornhill Magazine* was written

115

with no intention of trying really to investigate this subject, but to deride the notion that worked-stone objects have fallen from the sky. A writer in the *Amer. Jour. Sci.*, 1–21–325, read this paper and thinks it remarkable "that any man of ordinary reasoning powers should write a paper to prove that thunderbolts do not exist".

I confess that we're a little flattered by that.

Over and over:

"It is scarcely necessary to the intelligent reader that thunderstones are a myth."

We contend that there is a misuse of a word here: we admit that only we are intelligent upon this subject, if by intelligence is meant the inquiry of inequilibrium, and that all other intellection is only mechanical reflex—of course that intelligence, too, is mechanical, but less orderly and confined: less obviously mechanical—that as an acceptance of ours becomes firmer and firmer-established, we pass from the state of intelligent to reflexes in ruts. An odd thing is that intelligence is usually supposed to be creditable. It may be in the sense that it is mental activity trying to find out, but it is confession of ignorance. The bees, the theologians, the dogmatic scientists are the intellectual aristocrats. The rest of us are plebeians, not yet graduated to Nirvana, or to the instinctive and suave as differentiated from the intelligent and crude.

Blinkenberg gives many instances of the superstition of "thunderstones" which flourishes only where mentality is in a lamentable state—or universally. In Malacca, Sumatra, and Java, natives say that stone axes have often been found under trees that have been struck by lightning. Blinkenberg does not dispute this, but says it is coincidence: that the axes were of course upon the ground in the first place: that the natives jumped to the conclusion that these carved stones had fallen in or with lightning. In Central Africa, it is said that often have wedge-shaped, highly polished objects of stone, described as "axes", been found sticking in trees that have been struck by lightning—or by what seemed to be lightning. The natives rather like the unscientific persons of Memphis, Tenn., when they saw snakes after a storm, jumped to the conclusion that the "axes" had not always been sticking in the trees. Livingstone (*Last Journal*, pages 83, 89, 442, 448) says that he had never heard of stone implements used by natives of

116

Africa. A writer in the *Report of the Smithsonian Institution*, 1877–308, says that there are a few.

That they are said, by the natives, to have fallen in thunderstorms.

As to luminosity, it is lamentable acceptance that bodies falling through this earth's atmosphere, if not warmed even, often fall with a brilliant light, looking like flashes of lightning. This matter seems important: we'll take it up later, with data.

In Prussia, two stone axes were found in the trunks of trees, one under the bark. (Blinkenberg. *Thunder Weapons*, p. 100.)

The finders jumped to the conclusion that the axes had fallen there.

Another stone axe—or wedge-shaped object of worked stone—said to have been found in a tree that had been struck by something that looked like lightning. (*Thunder Weapons*, p. 71.)

The finder jumped to the conclusion.

Story told by Blinkenberg, of a woman, who lived near Kulsbjaergene, Sweden, who found a flint near an old willow—"near her house". I emphasize "near her house" because that means familiar ground. The willow had been split by something.

She jumped.

Cow killed by lightning, or by what looked like lightning (Isle of Sark, near Guernsey). The peasant who owned the cow dug up the ground at the spot and found a small greenstone "axe". Blinkenberg says that he jumped to the conclusion that it was this object that had fallen luminously, killing the cow.

Reliquary, 1867–208:

A flint axe found by a farmer, after a severe storm—described as a "fearful storm"—by a signal staff, which had been split by something. I should say that nearness to a signal staff may be considered familiar ground.

Whether he jumped, or arrived at the conclusion by more leisurely process, the farmer thought that the flint object had fallen in the storm.

In this instance we have a lamentable scientist with us. It's impossible to have positive difference between orthodoxy and heresy: somewhere there must be a merging into each other, or an overlapping. Nevertheless, upon such a subject as this, it does seem a little shocking. In most works upon meteorites, the peculiar, sulphurous odour of things that fall from the sky is mentioned. Sir John Evans (*Stone Implements*, p. 57) says—with extraordinary reasoning powers, if he could never have thought such a thing with ordinary reasoning powers—that this flint object "proved to have been the bolt, by its peculiar smell when broken".

If it did so prove to be, that settles the whole subject. If we prove that only one object of worked stone has fallen from the sky, all piling up of further reports is unnecessary. However, we have already taken the stand that nothing settles anything; that the disputes of ancient Greece are no nearer solution now than they were several thousand years ago—all because, in a positive sense, there is nothing to prove or solve or settle. Our subject is to be more nearly real than our opponents. Wideness is an aspect of the Universal. We go on widely. According to us the fat man is nearer godliness than is the thin man. Eat, drink, and approximate to the Positive Absolute. Beware of negativeness, by which we mean indigestion.

The vast majority of "thunderstorms" are described as "axes", but Meunier (*La Nature*, 1892-2-381) tells of one that was in his possession; said to have fallen at Ghardia, Algeria, contrasting "profoundment" (pear-shaped) with the angular outlines of ordinary meteorites. The conventional explanation that it had been formed as a drop of molten matter from a larger body seems reasonable to me; but with less agreeableness I note its fall in a thunderstorm, the datum that turns the orthodox meteorologist pale with rage, or induces a slight elevation of his eyebrows, if you mention it to him.

Meunier tells of another "thunderstone" said to have fallen in North Africa. Meunier, too, is a little lamentable here: he quotes a soldier of experience that such objects fall most frequently in the deserts of Africa.

Rather miscellaneous now:

"Thunderstone" said to have fallen in London, April 1876: weight about eight pounds: no particulars as to shape (Timb's *Year Book*, 1877-246).

"Thunderstone" said to have fallen at Cardiff, Sept, 26, 1916 (London *Times*, Sept. 28, 1916). According to *Nature*, 98-95, it was coincidence; only a lightning flash had been seen.

Stone that fell in a storm, near St. Albans, England: accepted by the Museum of St. Albans; said, at the British Museum, not to be of "true meteoritic material". (*Nature*, 80–34.)

London *Times*, April 26, 1876:

That, April 20, 1876, near Wolverhampton, fell a mass of meteoritic iron during a heavy fall of rain. An account of this phenomenon in *Nature*, 14–272, by H. S. Maskelyne, who accepts it as authentic. Also, see *Nature*, 13–531.

For three other instances, see the *Scientific American*, 47, 194; 52–83; 68–325.

As to wedge-shape larger than could very well be called an "axe":

Nature, 30–300:

That, May 27, 1884, at Tysnas, Norway, a meteorite had fallen: that the turf was torn up at the spot where the object had been supposed to have fallen; that two days later "a very peculiar stone" was found near by. The description is—"In shape and size very like the fourth part of a large Stilton cheese."

It is our acceptance that many objects and different substances have been brought down by atmospheric disturbance from what—only as a matter of convenience now, and until we have more data—we call the Super-Sargasso Sea; however, our chief interest is in objects that have been shaped by means similar to human handicraft.

Description of the "thunderstones" of Burma (*Proc. Asiatic Soc. of Bengal*, 1869–183): said to be of a kind of stone unlike any other found in Burma; called "thunderbolts" by the natives. I think there's a good deal of meaning in such expressions as "unlike any other found in Burma"—but that if they had anything more definite, there would have been unpleasant consequences to writers in the nineteenth century.

More about the "thunderstones" of Burma, in the *Proc. Soc. Antiqu. of London*, 2–3–97. One of them, described as an "adze", was exhibited by Captain Duff, who wrote that there was no stone like it in its neighbourhood.

Of course it may not be very convincing to say that because a stone is unlike neighbouring stones it had foreign origin—also we read it is a kind of plagiarism: we got it from the geologists, who demonstrate by this reasoning the foreign origin of erratics. We fear we're a little gross and scientific at times.

But it's my acceptance that a great deal of scientific literature be read between the lines. It's not everyone who has the lamentableness of a Sir John Evans. Just as a great deal of Voltaire's meaning was inter-linear, we suspect that a Captain Duff merely hints rather than risk having a Prof. Lawrence Smith fly at him and call him "a half–insane man". Whatever Captain Duff's meaning may have and whether he smiled like a Voltaire when he wrote it, Captain Duff writes of "the extremely soft nature of the stone, rendering it equally useless as an offensive weapon".

Story, by a correspondent, in *Nature*, 34–53, of a Malay, of "considerable social standing"—and one thing about our data is that, damned though they be, they do so often bring us awfully good company—who knew of a tree that had been struck, about a month before, by something in a thunderstorm. He searched among the roots of this tree and found a "thunderstone". Not said whether he jumped or leaped to the conclusion that it had fallen: process likely to be more leisurely in tropical countries. Also I'm afraid his way of reasoning was not very original: just so were fragments of the Bath-furnace meteorite, accepted by orthodoxy, discovered.

We shall now have an unusual experience. We shall read of some reports of extraordinary circumstances that were investigated by him, but that his phenomena occupied a position approximating higher to real investigation than to utter neglect. Over and over we read of extraordinary occurrences—no discussion; not even a comment afterward findable; mere mention occasionally—burial and damnation.

The extraordinary and how quickly it is hidden away.

Burial and damnation, or the obscurity of the conspicuous.

We did read of a man who, in the matter of snails, did travel some distance to assure himself of something that he had suspected in advance; and we remember Prof. Hitchcock, who had only to smite Amherst with the wand of his botanical knowledge, and lo! two fungi sprang up before night; and we did read of Dr. Gray and his thousands of fishes from one pailful of water—but these instances stand out; more frequently there was no "investigation". We now have a good many reported occurrences that were "investi-gated". Of things said to have fallen from the sky, we make, in the usual scientific way, two divisions: miscellaneous objects and substances, and symmetric objects attributable to beings like human beings, sub-dividing into—wedges, spheres, and discs.

Jour. Roy. Met. Soc., 14–207:

That, July 2, 1866, a correspondent to a London news-paper wrote that something had fallen from the sky, during a thunderstorm of June 30, 1866, at Notting Hill. Mr. G. T. Symons, of *Symons' Meteorological Magazine*, investi-gated, about as fairly, and with about as unprejudiced a mind, as anything ever has been investigated.

He says that the object was nothing but a lump of coal: that next door to the home of the correspondent coal had been unloaded the day before. With the uncanny wisdom of the stranger upon unfamiliar ground that we have noted before, Mr. Symons saw that the coal reported to have fallen from the sky, and the coal unloaded more prosaically the day before, were identical. Persons in the neighbour-hood, unable to make this simple identification, had brought from the correspondent pieces of the object reported to have fallen from the sky. As to credulity, I know of no limits for it—but when it comes to paying out money for credulity—oh, no standards to judge by, of course—just the same—

The trouble with efficiency is that it will merge away into excess. With what seems to me to be super-abundance of convincingness, Mr. Symons then lugs another character into his little comedy:

That it was all a hoax by a chemist's pupil, who had filled a capsule with an explosive, and "during the storm had thrown the burning mass into the gutter, so making an artificial thunderbolt".

Or even Shakespeare, with all his inartistry, did not lug in King Lear to make Hamlet complete.

Whether I'm lugging in something that has no special meaning myself, or not, I find that this storm of June 30, 1866, was peculiar. It is described in the London *Times*, July 2, 1866: that "during the storm, the sky in many places remained partially clear while hail and rain were falling." That may have more meaning when we take up the possible extra-mundane origin of some hailstones, especially if they fall from a cloudless sky. Mere suggestion, not worth much, that there may have been falls of extra-mundane substances, in London, June 30, 1866.

Clinkers, said to have fallen, during a storm, at Kilburn, July 5, 1877:

According to the *Kilburn Times*, July 7, 1877, quoted by Mr. Symons, a street had been "literally strewn," during the storm, with a mass of clinkers, estimated at about two bushels: sizes from that of a walnut to that of a man's hand—"pieces of the clinkers can be seen at the *Kilburn Times* office".

If these clinkers, or cinders, were refuse from one of the super-mercantile constructions from which coke, and coal and ashes occasionally fall to this earth, or, rather, to the Super-Sargasso Sea, from which dislodgement by tempests occurs, it is intermediatistic to accept that they must merge away somewhere with local phenomena of the scene of precipitation. If a red-hot stove should drop from a cloud into Broadway, someone would find that at about the time of the occurrence, a moving van had passed, and that the moving men had tired of the stove, or something—that it had not been really red-hot, but had been rouged instead of blacked, by some absentminded housekeeper. Compared with some of the scientific explanations that we have encountered, there's considerable restraint, I think, in that one.

Mr. Symons learned that in the same street—he emphasizes that it was a short street—there was a fire-engine station. I had such an impression of him hustling and bustling around at Notting Hill, searching cellars until he found one with newly arrived coal in it; ringing door bells, exciting a whole neighbourhood, calling up to second-storey windows, stopping people in the streets, hotter and hotter on the trail of a wretched impostor of a chemist's pupil. After his efficiency at Notting Hill, we'd expect to hear that he went to the station, and—something like this:

"It is said that clinkers fell, in your street, at about ten minutes past four o'clock, afternoon of July fifth. Will you look over your records and tell me where your engine was at about ten minutes past four, July fifth?"

Mr. Symons says:

"I think that most probably they had been raked out of the steam fire-engine."

June 20, 1880, it was reported that a "thunderstone" had struck the house at 180 Oakley Street, Chelsea, falling down the chimney, into the kitchen grate.

Mr. Symons investigated.

He describes the "thunderstone" as an "agglomeration of brick, soot, unburned coal, and cinder".

He says that, in his opinion, lightning had flashed down the chimney, and had fused some of the brick of it.

He does think it remarkable that the lightning did not then scatter the contents of the grate, which were disturbed only as if a heavy body had fallen. If we admit that climbing up the chimney to find out is too rigorous a requirement for a man who may have been large, dignified and subject to expansions, the only unreasonableness we find in what he says—as judged by our more modern outlook, is:

"I suppose that no one would suggest that bricks are manufactured in the atmosphere."

Sounds a little unreasonable to us, because it is so of the positivistic spirit of former times, when it was not so obvious that the highest incredibility and laughability must merge away with the "proper"—as the *Sci. Am. Sup.* would say. The preposterous is always interpretable in terms of the "proper," with which it must be continuous—or—clay-like masses such as have fallen from the sky—tremendous heat generated by their velocity—they bake—bricks.

We begin to suspect that Mr. Symons exhausted himself at Notting Hill. It's a warning to efficiency-fanatics.

Then the instance of three lumps of earthy matter, found upon a well-frequented path, after a thunderstorm, at Reading, July 3, 1883. There are so many records of the fall of earthy matter from the sky that it would seem almost uncanny to find resistance here, were we not so accustomed to the uncompromising stands of orthodoxy—which, in our metaphysics, represent good, as attempts, but evil in their insufficiency. If I thought it necessary, I'd list one hundred and fifty instances of earth matter said to have fallen from the sky. It is his antagonism to atmospheric disturbance associated with the fall of things from the sky that blinds and hypnotizes a Mr. Symons here. This especial Mr. Symons rejects the Reading substance because it was not "of true meteoritic material". It's uncanny—or it's not uncanny at all, but universal—if you don't take something for a standard of opinion, you can't have any opinion at all: but, if you do take a standard, in some of its applications it must be preposterous. The carbonaceous meteorites, which are unquestioned—though avoided, as we have seen—by orthodoxy, are more glaringly of untrue meteoritic material than was this substance of Reading. Mr. Symons says that these three lumps were upon the ground "in the first place".

Whether these data are worth preserving or not, I think that the appeal that this especial Mr. Symons makes is worthy of a place in the museum we're writing. He argues against belief in all external origins "for our credit as Englishmen". He is a patriot, but I think that these foreigners had a small chance "in the first place" for hospitality from him.

Then comes a "small lump of iron (two inches in diameter)" said to have fallen, during a thunderstorm, at Brixton, Aug. 17, 1887. Mr. Symons says: "At present I cannot trace it."

He was at his best at Notting Hill: there's been a marked falling off in his later manner:

In the London *Times*, Feb. 1, 1888, it is said that a roundish object of iron had been found, "after a violent thunderstorm", in a garden at Brixton, Aug. 17, 1887. It was analysed by a chemist, who could not identify it as true meteoric material. Whether a product of workmanship like human workmanship or not, this object is described as an oblate spheroid, about two inches across its major diameter. The chemist's name and address are given: Mr. James J. Morgan: Ebbw Vale.

Garden—familiar ground—I suppose that in Mr. Symons' opinion this symmetric object had been upon the ground "in the first place", though he neglects to say this. But we do note that he describes this object as a "lump" which does not suggest the spheroidal or symmetric. It is our notion that the word "lump" was, because of its meaning of amorphousness, used purposely to have the next datum stand alone, remote, without similars. If Mr. Symons had said that there had been a report of another round object that had fallen from the sky, his readers would be attracted by an agreement. He distracts his readers by describing in terms of the unprecedented—

"Iron cannon ball."

It was found in a manure heap, in Sussex, after a thunderstorm.

However, Mr. Symons argues pretty reasonably, it seems to me, that, given a cannon ball in a manure heap, in the first place, lightning might be attracted by it, and, if seen to strike there, the untutored mind, or mentality below the average, would leap or jump, or proceed with less celerity, to the conclusion that the iron object had fallen.

Except that—if every farmer isn't upon very familiar ground—or if every farmer doesn't know his own manure heap as well as Mr. Symons knew his writing desk—

Then comes the instance of a man, his wife, and his three daughters, at Casterton, Westmoreland, who were looking out at their lawn, during a thunderstorm, when they "considered", as Mr. Symons expresses it, that they saw a stone fall from the sky, kill a sheep, and bury itself in the ground.

They dug.

They found a stone ball.

Symons:

Coincidence. It had been there in the first place.

This object was exhibited at a meeting of the Royal Meteorological Society by Mr. C. Carus-Wilson. It is described in the *Journal*'s list of exhibits as a "sandstone" ball. It is described as "sandstone" by Mr. Symons.

Now a round piece of sandstone may be almost anywhere in the ground—in the first place—but, by our more or less discreditable habit of prying and snooping, we find that this object was rather more complex and of material less commonplace. In snooping through *Knowledge*, Oct. 9, 1885, we read that this "thunderstone" was in the possession of Mr. C. Carus-Wilson, who tells the story of the witness and his family—the sheep killed, the burial of something in the earth, the digging, and the finding. Mr. C. Carus-Wilson describes the object as a ball of hard, ferruginous quartzite, about the size of a coconut, weight about 12 pounds. Whether we're feeling around for significance or not, there is a suggestion not only of symmetry but of structure in this object: it had an external shell, separated from a loose nucleus. Mr. Carus-Wilson attributes this cleavage to unequal cooling of the mass.

My own notion is that there is very little deliberate misrepresentation in the writing of scientific men: that they are quite as guiltless in intent as are other hypnotic subjects.

Such a victim of induced belief reads of a stone ball said to have fallen from the sky. Mechanically in his mind arise impressions of globular lumps, or nodules, of sandstone, which are common almost everywhere. He assimilates the reported fall with his impression of objects in the ground, in the first place. To an intermediatist, the phenomena of intellection are only phenomena of universal process localized in human minds. The process called "explanation" is only a local aspect of universal assimilation. It looks like materialism: but the intermediatist holds that interpretation of the immaterial, as it is called, in terms of the material, as it is called, is no more rational than interpretation of the "material" in terms of the "immaterial": that there is in quasi-existence neither the material nor the immaterial, but approximations one way or the other. But so hypnotic quasi-reasons: that globular lumps of sandstone are common. Whether he jumps or leaps, or whether only the frowsy and baseborn are so athletic, his is the impression, by assimilation, that this especial object is a ball of sandstone. Or human mentality: its inhabitants as conveniences. It may be that Mr. Symons' paper was written before this object was exhibited to the members of the Society, and with the charity with which, for the sake of diversity, we intersperse our malices, we are willing to accept that he "investigated" something that he had never seen. But whoever listed this object was uncareful: it is listed as "sandstone".

We're making excuses for them.

Really—as it were—you know, we're not quite so damned as we were.

One does not apologize for the gods and at the same time feel quite utterly prostrate before them.

If this were a real existence, and all of us real persons, with real standards to judge by, I'm afraid we'd have to be a little more severe with some of these Mr. Symonses. As it is, of course, seriousness seems out of place.

We note an amusing little touch in the indefinite allusion to "a man", who with his un-named family, had "considered" that he had seen a stone fall. The "man" was the Rev. W. Carus-Wilson, who was well-known in his day.

The next instance was reported by W. B. Tripp, F.R.M.S.—that, during a thunderstorm, a farmer had seen the ground in front of him ploughed up by something that was luminous.

Dug.

Bronze axe.

My own notion is that an expedition to the North Pole could not be so urgent as that representative scientists should have gone to that farmer and there spent a summer studying this one reported occurrence. As it is—unnamed farmer—somewhere—no date. The thing must stay damned.

Another specimen for our museum is a comment in *Nature* upon these objects: that they are "of an amusing character, thus clearly showing that they were of terrestrial, and not a celestial, character." Just why celestiality, or that of it which, too, is only of Intermediateness should not be quite as amusing as terrestriality is beyond our reasoning powers, which we have agreed are not ordinary. Of course there is nothing amusing about wedges and spheres at all—or Archimedes and Euclid are humorists. It is that they were described derisively. If you'd like a little specimen of the standardization of orthodox opinion—

Amer. Met. Jour., 4–589:

"They are of an amusing character, thus clearly showing that they were of a terrestrial and not a celestial character."

I'm sure—not positively, of course—that we've tried to be as easy going and lenient with Mr. Symons as his obviously scientific performance would permit. Of course it may be that sub-consciously we were prejudiced against him, instinctively, classing him with St. Augustine, Darwin, St. Jerome, and Lyell. As to the "thunderstones", I think that he investigated them mostly "for the credit of Englishmen," or in the spirit of the Royal Krakatoa Committee, or about as the commission for the French Academy investigated meteorites. According to a writer in *Knowledge*, 5–418, the Krakatoa Committee attempted not in the least to prove what had caused the atmospheric effects of 1883, but to prove—that Krakatoa did it.

Altogether I should think that the following quotation should be enlightening to anyone who still thinks that these occurrences were investigated not to support an opinion formed in advance:

In opening his paper, Mr. Symons said that he undertook his investigation as to the existence of "thunderstones", or "thunderbolts" as he calls them—"feeling certain that there was a weak point somewhere, in as much as 'thunderbolts' have no existence".

We have another instance of the reported fall of a "cannon ball". It occurred prior to Mr. Symons' investigations, but is not mentioned by him. It was investigated, however. In the *Proc. Roy. Soc. Edin.*, 3–147, is the report of a "thunderstone", "supposed to have fallen in Hampshire, Sept., 1852". It was an iron cannon ball, or it was a "large nodule of iron pyrites or bisulphuret of iron". No one had seen it fall. It had been noticed, upon a garden path, for the first time, after a thunderstorm. It was only a "supposed" thing, because—"It had not the character of any known meteorite."

In the London *Times*, Sept. 16, 1852, appears a letter from Mr. George E. Bailey, a chemist of Andover, Hants. He says that, in a very heavy thunderstorm, of the first week of September, 1852, this iron object had fallen in the garden of Mr. Robert Dowling, of Andover; that it had fallen upon a path "within six yards of the house". It had been picked up "immediately" after the storm by Mrs. Dowling. It was about the size of a cricket ball: weight four pounds. No one had seen it fall. In the *Times*, Sept. 15, 1852, there is an account of this thunderstorm, which was of unusual violence.

There are some other data relative to the ball of quartz of Westmoreland. They're poor things. There's so little to them that they look like ghosts of the damned. However, ghosts, when multiplied, take on what is called substantiality—if the solidest thing conceivable, in quasi-existence, is only concentrated phantomosity. It is not only that there have been other reports of quartz that has fallen from the sky; there is another agreement. The round quartz object of Westmoreland, if broken open and separated from its loose nucleus, would be a round, hollow, quartz object. My pseudo-position is that two reports of similar extraordinary occurrence, one from England and one from Canada—are interesting.

Proc. Canadian Institute, 3–7–8:

That, at the meeting of the Institute, of Dec. 1, 1888, one of the members, Mr. J. A. Livingstone, exhibited a globular quartz body which he asserted had fallen from the sky. It had been split open. It was hollow.

128

But the other members of the Institute decided that the object was spurious, because it was not of "true meteoritic material".

No date; no place mentioned; we note the suggestion that it was only a geode, which had been upon the ground in the first place. Its crystalline lining was geode-like.

Quartz is upon the "index prohibitory" of Science. A monk who would read Darwin would sin no more than would a scientist who would admit that, except by the "up and down" process, quartz has never fallen from the sky—but Continuity: it is not excommunicated if part of or incorporated in a baptized meteorite—St. Catherine's of Mexico, I think. It's as epicurean a distinction as any ever made by theologians. Fassig lists a quartz pebble, found in a hailstone (*Bibliography*, part 2–355). "Up and down," of course. Another object of quartzite was reported to have fallen, in the autumn of 1880, at Schroon Lake, N.Y.—said in the *Scientific American*, 43–272, to be a fraud—it was not—the usual. About the first of May, 1899, the newspapers published a story of a "snow-white" meteorite that had fallen, at Vincennes, Indiana. The Editor of the *Monthly Weather Review* (issue of April, 1899) requested the local observer, at Vincennes, to investigate. The Editor says that the thing was only a fragment of a quartz boulder. He says that anyone with at least a public school education should know better than to write that quartz has ever fallen from the sky.

Notes and Queries, 2–8–92:

That, in the Leyden Museum of Antiquities, there is a disc of quartz: six centimetres, by five millimetres by about five centimetres; said to have fallen upon a plantation in the Dutch West Indies, after a meteoric explosion.

Bricks.

I think this is a vice we're writing. I recommend it to those who have hankered for a new sin. At first some of our data were of so frightful or ridiculous mien as to be hated, or eyebrowed, was only to be seen. Then some pity crept in? I think that we can now embrace bricks.

The baked-clay idea was right in its place, but it rather lacks distinction, I think. With our minds upon the concrete boats that have been building terrestrially lately, and thinking of wrecks that may occur to some of them, and of a new material for the deep-sea fishes to disregard—

Object that fell at Richland, South Carolina—yellow to grey—said to look like a piece of brick. (*Amer. Jour. Sci.*, 2–34–298.)

Pieces of "furnace-made brick" said to have fallen—in a hailstorm—at Padua, August, 1834. (*Edin. New Phil. Jour.*, 19–87.) The writer offered an explanation that started another convention: that the fragments of brick had been knocked from buildings by the hailstones. But there is here a concomitant that will be disagreeable to anyone who may have been inclined to smile at the now digestible-enough notion that furnace-made bricks have fallen from the sky. It is that in some of the hailstones—two per cent of them—that were found with the pieces of brick, was a light greyish powder.

Monthly Notices of the Royal Astronomical Society, 335–365:

Padre Sechi explains that a stone said to have fallen, in a thunderstorm, at Supino, Italy, September, 1875, had been knocked from a roof.

Nature, 33–153:

That it had been reported that a good-sized stone, of form clearly artificial, had fallen at Naples, November, 1885. The stone was described by two professors of Naples, who had accepted it as inexplicable but veritable. They were visited by Dr. H. Johnstone-Lavis, the correspondent to *Nature*, whose investigations had convinced him that the object was a "shoemaker's lapstone".

Now to us of the initiated, or to us of the wider outlook, there is nothing incredible in the thought of shoemakers in other worlds—but I suspect that this characterization is tactical.

This object of worked stone, or this shoemaker's lapstone, was made of Vesuvian lava, Dr. Johnstone-Lavis thinks: most probably lava of the flow of 1631, from the La Scala quarries. We condemn "most probably" as bad positivism. As to the "men of position", who had accepted that this had fallen from the sky—"I have now obliged them to admit their mistake," says Dr. Johnstone-Lavis—or it's always the stranger in Naples who knows La Scala lava better than the natives know it.

Explanation:

That the thing had been knocked from, or thrown from, a roof.

As to attempt to trace the occurrence to any special roof—nothing said upon the subject. Or that Dr. Johnstone-Lavis called a carved stone a "lapstone", quite as Mr. Symons called a spherical object a "cannon ball": bent upon a discrediting incongruity:

Shoemaking and celestiality.

It is easy to say that axes, or wedge-shaped stones found on the ground, were there in the first place, and that it is only coincidence that lightning should strike near one—but the credibility of coincidence decreases as the square root of their volume, I think. Our massed instances speak too much of coincidences of coincidences. But the axes, or wedge-shaped objects that have been found in trees, are more difficult for orthodoxy. For instance, Arago accepts that such finds have occurred, but he argues that, if wedge-shaped stones have been found in tree trunks, so have toads been found in tree trunks—did the toads fall there?

Not at all bad for a hypnotic.

Of course, in our acceptance, the Irish are the Chosen People. It's because they are characteristically best in accord with the underlying essence of quasi-existence. M. Arago answers a question by asking another question. That's the only way a question can be answered in our Hibernian kind of existence.

Dr. Bodding argued with the natives of the Santal Paraganas, India, who said that cut and shaped stones had fallen from the sky, some of them lodging in tree trunks. Dr. Bodding, with orthodox notions of velocity of falling bodies, having missed, I suppose, some of the notes I have upon large hailstones, which, for size, have fallen with astonishingly low velocity argued that anything falling from the sky would be "smashed to atoms". He accepts that objects of worked stone have been found in tree trunks, but he explains:

That the Santals often steal trees, but do not chop them down in the usual way, because that would be to make too much noise: they insert stone wedges, and hammer them instead: then, if they should be caught, wedges would not be the evidence against them that axes would be.

Or that a scientific man can't be desperate and reasonable too.

Or that a pickpocket, for instance, is safe, though caught

with his hand in one's pocket, if he's gloved, say: because no court in the land would regard a gloved hand in the same way in which a bare hand would be regarded.

That there's nothing but intermediateness to the rational and the preposterous: that this status of our own ratiocinations is perceptible wherein they are upon the unfamiliar.

Dr. Bodding collected 50 of these shaped stones, said to have fallen from the sky, in the course of many years. He says that the Santals are a highly developed race, and for ages have not used stone implements—except in this one nefarious convenience to him.

All explanations are localizations. They fade away before the universal. It is difficult to express that black rains in England do not originate in the smoke of factories—less difficult to express that black rains of South Africa do not. We utter little stress upon the absurdity of Dr. Bodding's explanation, because, if anything's absurd everything's absurd, or, rather, has in it some degree or aspect of absurdity, and we've never had experience with any state except something somewhere between ultimate absurdity and final reasonableness. Our acceptance is that Dr. Bodding's elaborate explanations does not apply to cut-stone objects found in tree trunks in other lands: we accept that for the general, a local explanation is inadequate.

As to "thunderstones" not said to have fallen luminously, and not said to have been found sticking in trees, we are told by faithful hypnotics that astonished rustics come upon prehistoric axes that have been washed into sight by rains, and jump to the conclusion that the things have fallen from the sky. But simple rustics come upon many prehistoric things: scrapers, pottery, knives, hammers. We have no record of rusticity coming upon old pottery after a rain, reporting the fall of a bowl from the sky.

Just now, my own acceptance is that wedge-shaped stone objects, formed by means similar to human workmanship, have often fallen from the sky. Maybe there are messages upon them. My acceptance is that they have been called "axes" to discredit them: or the more familiar a term, the higher the incongruity with vague concepts of the vast, remote, tremendous, unknown.

In *Notes and Queries*, 2–892, a writer says that he had a "thunderstone", which he had brought from Jamaica. The description is of a wedge-shaped object; not of an axe:

"It shows no mark of having been attached to a handle."

Of ten "thunderstones", figured upon different pages in Blinkenberg's book, nine show no sign of ever having been attached to a handle: one is perforated.

But in a report by Dr. C. Leemans, Director of the Leyden Museum of Antiquities, objects, said by the Japanese to have fallen from the sky, are alluded to throughout as "wedges". In the *Archaeological Journal*, 11–118, in a paper upon the "Thunderstones" of Java, the objects are called "wedges" and not "axes".

Our notion is that rustics and savages call wedge-shaped objects that fall from the sky, "axes": that scientific man, when it suits their purposes, can resist temptations to prolixity and pedantry, and adopt the simple: that they can be intelligible when derisive.

All of which lands us in a confusion, worse, I think, than we were in before we so satisfactorily emerged from the distresses of—butter and blood and ink and paper and punk and silk. Now it's cannon balls and axes and discs—if a "lapstone" be a disc—it's a flat stone, at any rate.

A great many scientists are good impressionists: they snub the impertinences of details. Had he been of a coarse, grubbing nature, I think Dr. Bodding could never have so simply and beautifully explained the occurrence of stone wedges in tree trunks. But to a realist, the story would be something like this:

A man who needed a tree, in a land of jungles, where, for some unknown reason, everyone's very selfish with his trees, conceives that hammering stone wedges makes less noise than does the chopping of wood: he and his descendants, in a course of many years, cut down trees with wedges, and escape penalty, because it never occurs to a prosecutor that the head of an axe is a wedge.

The story is like every other attempted positivism—beautiful and complete, until we see what it excludes or disregards; whereupon it becomes the ugly and incomplete—but not absolutely, because there is probably something of what is called foundation for it. Perhaps a mentally incomplete Santal did once do something of the kind. Story told to Dr. Bodding: in the usual scientific way, he makes a dogma of an aberration.

Or we did have to utter a little stress upon this matter, after all. They're so hairy and attractive, these scientists of

133

the nineteenth century. We feel the zeal of a Sitting Bull when we think of their scalps. We shall have to have an expression of our own upon this confusing subject. We have expressions: we don't call them explanations: we've discarded explanations with beliefs. Though everyone who scalps is, in the oneness of allness, himself likely to be scalped, there is such a discourtesy to an enemy as the wearing of wigs.

Cannon balls and wedges, and what may they mean?

Bombardments of this earth—

Attempts to communicate—

Or visitors to this earth, long ago—explorers from the moon—taking back with them, as curiosities, perhaps, implements of this earth's prehistoric inhabitants—a wreck—a cargo of such things held for ages in suspension in the Super-Sargasso Sea—falling, or shaken, down occasionally by storms—

But, by preponderance of description, we cannot accept that "thunderstorms" ever were attached to handles, or are prehistoric axes—

As to attempts to communicate with this earth by means of wedge-shaped objects especially adapted to the penetration of vast, gelatinous areas spread around this earth—

In the *Proc. Roy. Irish Acad.*, 9–337, there is an account of a stone wedge that fell from the sky, near Cashel, Tipperary, Aug. 2, 1865. The phenomenon is not questioned, but the orthodox preference is to call it, not axe-like, nor wedge-shaped, but "pyramidal." For data of other pyramidal stones said to have fallen from the sky, see *Rept. Brit. Assoc.*, 1861–34. One fell at Segowolee, India, March 6, 1853. Of the object that fell at Cashel, Dr. Haughton says in the *Proceedings*: "A singular feature is observable in this stone, that I have never seen in any other:—the rounded edges of the pyramid are sharply marked by lines on the black crust, as perfect as if made by a ruler." Dr. Haughton's idea is that the marks may have been made by "some peculiar tension in the cooling". It must have been very peculiar, if in all aerolites not wedge-shaped, no such phenomenon had ever been observed. It merges away with one or two instances known, after Dr. Haughton's time of seeing stratification in meteorites. Stratification in meteorites, however, is denied by the faithful.

I begin to suspect something else.

A whopper is coming.

Later it will be as reasonable, by familiarity, as anything else ever said.

If someone should study the stone of Cashel, as Champollion studied the Rosetta stone, he might—or, rather, would inevitably—find meaning in those lines, and translate them into English—

Nevertheless I begin to suspect something else: something more subtle and esoteric than graven characters upon stones that have fallen from the sky, in attempts to communicate. The notion that other worlds are attempting to communicate with this world is wide-spread: my own notion is that it is not an attempt at all—that it was achievement centuries ago.

I should like to send out a report that a "thunderstone" had fallen, say, somewhere in New Hampshire—

And keep track of every person who came to examine that stone—trace down his affiliations—keep track of him—

Then send out a report that a "thunderstone" had fallen at Stockholm, say—

Would one of the persons who had gone to New Hampshire, be met again in Stockholm? But—when if he had no anthropological, lapidarian, or meteorological affiliations—but did not belong to a secret society—

It is only a dawning credulity.

Of the three forms of symmetric objects that have, or haven't, fallen from the sky, it seems to me that the disc is the most striking. So far, in this respect, we have been at our worst—possibly that's pretty bad—but "lapstones" are likely to be of considerable variety of form, and something that is said to have fallen at some time somewhere in the Dutch West Indies is profoundly of the unchosen.

Now we shall have something that is high up in the castes of the accursed:

Comptes Rendus, 1887–182:

That, upon June 20, 1887, in a "violent storm"—two months before the reported fall of the symmetric iron of Brixton—a small stone had fallen from the sky at Tarbes, France: 13 millimetres in diameter; five millimetres thick; weight two grammes. Reported to the French Academy by M. Sudre, professor of the Normal School, Tarbes.

This time the old convenience "there in the first place" is too greatly resisted—the stone was covered with ice.

This object had been cut and shaped by means similar to human hands and human mentality. It was a disc of worked stone—"très regulier". "Il a été assurement travaillé."

There's not a word as to any known whirlwind anywhere: nothing of other objects or débris that fell at or or near this date, in France. The thing had fallen alone. But as mechanically as any part of a machine responds to its stimulus, the explanation appears in *Comptes Rendus* that this stone had been raised by a whirlwind and then flung down.

It may be that in the whole nineteenth century no event more important than this occurred. In *La Nature*, 1887, and in *L'Année Scientifique*, 1887, this occurrence is noted. It is mentioned in one of the summer numbers of *Nature*, 1887. Fassig lists a paper upon it in the *Annuaire de Soc., Met.*, 1887.

Not a word of discussion.

Not a subsequent mention can I find.

Our own expression:

What matters is how we, the French Academy, or the Salvation Army may explain?

A disc of worked stone fell from the sky, at Tarbes, France, June 20, 1887.

Chapter 9

My own pseudo-conclusion:

That we've been damned by giants sound asleep, or by great scientific principles and abstractions that cannot realize themselves: that little harlots have visited their caprices upon us; that clowns, with buckets of water from which they tend to cast thousands of good-sized fishes have anathematized us for laughing disrespectfully, because, as with all clowns, underlying buffoonery is the desire to be taken seriously; that pale ignorances, presiding over microscopes by which they cannot distinguish flesh from nostoc or fishes' spawn or frogs' spawn, have visited upon us their wan solemnities. We've been damned by corpses and skeletons and mummies, which twitch and totter with pseudo-life derived from conveniences.

Or there is only hypnosis. The accursed are those who admit they're the accursed.

If we be more nearly real we are reasons arraigned before a jury of dream-phantasms.

Of all meteorites in museums, very few were seen to fall. It is considered sufficient grounds for admission if specimens can't be accounted for in any way other than that they fell from the sky—as if in the haze of uncertainty that surrounds all things, or that is the essence of everything, or in the merging away of everything into something else, there could be anything that could be accounted for in only one way. The scientist and the theologian reason that if something can be accounted for in only one way, it is accounted for in that way—or logic would be logical, if the conditions that it imposes, but, of course, does not insist upon, could anywhere be found in quasi-existence. In our acceptance, logic, science, art, religion are, in our "existence", premonitions of a coming awakening, like dawning awareness of surroundings in the mind of a dreamer.

Any old chunk of metal that measures up to the standard of "true meteoritic material" is admitted by the museums. It may seem incredible that modern curators still have this delusion, but we suspect that the date on one's morning newspaper hasn't much to do with one's modernity all day long. In reading Fletcher's catalogue, for instance, we learn that some of the best-known meteorites were "found in draining a field"—"found in the making of a road"—"turned up by the plough" occurs a dozen times. Someone fishing in Lake Okeechobee, brought up an object in his fishing net. No meteorite had ever been seen to fall near it. The U.S. National Museum accepts it.

If we have accepted only one of the data of "untrue meteoritic material"—one instance of "carbonaceous" matter—if it be too difficult to utter the word "coal"—we see that in this inclusion-exclusion, as in every other means of forming an opinion, false inclusion and false exclusion have been practised by curators of museums.

There is something of ultra-pathos—of cosmic sadness—in this universal search for a standard, and in belief that one has been revealed by either inspiration or analysis, than the dogged clinging to a poor sham of a thing long after its insufficiency has been shown—or renewed hope and search for the special that can be true, or for something local that could also be universal. It's as if "true meteoritic material" were a "rock of ages" to some scientific men. They cling. But clingers cannot hold out welcoming arms.

The only seemingly conclusive utterance, or seemingly substantial thing to cling to, is a product of dishonesty, ignorance, or fatigue. All sciences go back and back, until they're worn out with the process, or until mechanical reaction occurs: then they move forward—as it were. Then they become dogmatic, and take for bases, positions that were only points of exhaustion. So chemistry divided and sub-divided down to atoms; then, in the essential insecurity of all quasi-constructions, it built up a system, which, to anyone so obsessed by his own hypnoses that he is exempt to the chemist's hypnoses, is perceptibly enough and intellectual anaemia built upon infinitesimal debilities.

In *Science*, n.s., 31–298, E. D. Hovey, of the American Museum of Natural History, asserts or confesses that often have objects of material such as fossiliferous limestone and slag been sent to him. He says that these things have been accompanied by assurances that they have been seen to fall on lawns, on roads, in front of houses.

They are all excluded. They are not of true meteoritic material. They were on the ground in the first place. It is only by coincidence that lightning has struck, or that a real meteorite, which was unfindable, has struck near objects of slag and limestone.

Mr. Hovey says that the list might be extended indefinitely. That's a tantalizing suggestion of some very interesting stuff—

He says:

"But it is not worth while."

I'd like to know what strange, damned, excommunicated things have been sent to museums by persons who have felt convinced that they have seen what they may have seen, strongly enough to risk ridicule, to make up bundles, go to express offices, and write letters. I accept that over the door of every museum, into which such things enter, is written:

"Abandon Hope."

If a Mr. Symons mentions one instance of coal, or of slag or cinders, said to have fallen from the sky, we are not—except by association with the "carbonaceous" meteorites—strong in our impression that coal sometimes falls to this earth from coal-burning super-constructions up somewhere—

In *Comptes Rendus*, 91–197, M. Daubrée tells the same story. Our acceptance, then, is that other curators could tell this same story. Then the phantomosity of our impression substantiates proportionately to its multiplicity. M. Daubrée says that often have strange damned things been sent to the French museums, accompanied by assurances that they had been seen to fall from the sky. Especially to our interest, he mentions coal and slag.

Excluded.

Buried unnamed and undated in Science's potter's field.

I do not say that the data of the damned should have the same rights as the data of the saved. That would be justice. That would be of the Positive Absolute, and, though the ideal of, a violation of, the very essence of quasi-existence, wherein only to have the appearance of being is to express a preponderance of force one way or another—or inequilibrium, or inconsistency, or injustice.

Our acceptance is that the passing away of exclusionism is a phenomenon of the twentieth century: that gods of the twentieth century will sustain our notions be they ever so unwashed and frowsy. But, in our own expressions, we are limited, by the oneness of quasiness, to the very same methods by which orthodoxy established and maintains its now sleek, suave preposterousnesses. At any rate, though we are inspired by an especial subtle essence—or imponderable, I think—that pervades the twentieth century, we have not the superstition that we are offering anything as a positive fact. Rather often we have not the delusion that we're any less superstitious and credulous than any logician, savage, curator, or rustic.

An orthodox demonstration, in terms of which we shall have some heresies, is that if things found in coal could have got there only by falling there—they fell there.

So, in the *Manchester Lit. and Phil. Soc. Mems.*, 2–9–306, it is argued that certain roundish stones that have been found in coal are "fossil aerolites"; that they had fallen from the sky, ages ago, when the coal was soft, because the coal had closed around them, showing no sign of entrance.

Proc. Soc. of Antiz. of Scotland, 1–1–121:

That, in a lump of coal, from a mine in Scotland, an iron instrument had been found—

"The interest attaching to this singular relic arises from the fact of its having been found in the heart of a piece of coal, seven feet under the surface."

If we accept that this object of iron was of workmanship beyond the means and skill of the primitive men who may have lived in Scotland when coal was forming there—

"The instrument was considered to be modern."

That our expression has more of realness, or higher approximation to realness, than has the attempt to explain that is made in the *Proceedings:*

That in modern times someone may have bored for coal, and that his drill may have broken off in the coal it had penetrated.

Why he should have abandoned such easily accessible coal, I don't know. The important point is that there was no sign of boring: that this instrument was in a lump of coal that had closed around it so that its presence was not suspected, until the lump of coal was broken.

No mention can I find of this damned thing in any other publication. Of course there is an alternative here: the thing may not have fallen from the sky: if in coal-forming times, in Scotland, there were, indigenous to this earth, no men capable of making such an iron instrument, it may have been left behind by visitors from other worlds.

In an extraordinary approximation to fairness and justice, which is permitted to us, because we are quite as desirous to make acceptable we note:

That in *Notes and Queries*, 11-1-408, there is an account of an ancient copper seal, about the size of a penny, found in chalk, at a depth of from five to six feet, near Brednenstone, England. The design upon it is said to be of a monk kneeling before a virgin and child: a legend upon the margin is said to be: "St. Jordanis Monachi Spaldingie."

I don't know about that. It looks very desirable—undesirable to us.

There's a wretch of an ultra-frowsy thing in the *Scientific American*, 7-298, which we condemn ourselves, if somewhere, because of the oneness of allness, the damn must also be the damning. It's a newspaper story: that about the first of June, 1851, a powerful blast, near Dorchester, Mass., cast out from a bed of solid rock a bell-shaped vessel of an unknown metal: floral designs inlaid with silver; "art of some cunning workman". The opinion of the Editor of the *Scientific American* is that the thing had been made by Tubal Cain, who was the first inhabitant of Dorchester. Though I fear that this is a little arbitrary, I am not disposed to fly rabidly at every scientific opinion.

Nature, 35-36:

A block of metal found in coal, in Austria, 1885. It is now in the Salzburg museum.

This time we have another expression. Usually our intermediatist attack upon provincial positivism is: Science, in its attempted positivism take something such as "true meteoritic material" as a standard of judgment; but carbonaceous matter, except for its relative infrequency, is just as veritable a standard of judgment; carbonaceous matter merges away into such a variety of organic substances, that all standards are reduced to indistinguishability: if, then, there is no real standard against us, there is no real resistance to our own acceptances. Now our intermediatism is: Science takes "true meteoritic material" as a standard of admission; but now we have an instance that quite as truly makes "true meteoritic material" a standard of exclusion; or, then, a thing that denies itself is no real resistance to our own acceptances—this depending upon whether we have a datum of something of "true meteoritic material" that orthodoxy can never accept fell from the sky.

We're a little involved here. Our own acceptance is upon a carved geometric thing that, if found in a very old deposit, antedates human life, except, perhaps, very primitive human life, as an indigenous product of this earth: but we're quite as much interested in the dilemma it made for the faithful.

It is of "true meteoritic material". In *L'Astronomie*. 1887–114, it is said that, though so geometric, its phenomena so characteristic of meteorites exclude the idea that it was the work of man.

As to the deposit—Tertiary coal.

Composition—iron, carbon, and a small quantity of nickel,

It has the pitted surface that is supposed by the faithful to be characteristic of meteorites.

For a full account of this subject, see *Comptes Rendus*, 103–702. The scientists who examined it could reach no agreement. They bifurcated: then a compromise was suggested; but the compromise is a product of disregard:

That it was of true meteoritic material, and had not been shaped by man;

That it was not of true meteoritic material, but telluric iron that had been shaped by man;

That it was true meteoritic material that had fallen from the sky, but had been shaped by man, after its fall.

The data, one or more of which must be disregarded by each of these three explanations, are: "true meteoritic material" and surface markings of meteorites; geometric form; presence in an ancient deposit; material as hard as steel; absence upon this earth, in Tertiary times, of men who could work in material as hard as steel. It is said that, though of "true meteoritic material", this object is virtually a steel object.

St. Augustine, with his orthodoxy, was never in—well, very much worse—difficulties than are the faithful here. By due disregard of a datum or so, our own acceptance that it was a steel object that had fallen from the sky to this earth, in Tertiary times, is not forced upon one. We offer ours as the only synthetic expression. For instance, in *Science Gossip*, 1887–58, it is described as a meteorite; in this account there is nothing alarming to the pious, because though everything else is told, its geometric form is not mentioned.

It's a cube. There is a deep incision all around it. Of its faces, two that are opposite are rounded.

Though I accept that our own expression can only rather approximate to Truth by the wideness of its inclusions and because it seems, of four attempts, to represent the only complete synthesis, and can be nullified or greatly modified by data that we, too, have somewhere disregarded, the only means of nullification that I can think of would be demonstration that this object is a mass of iron pyrites, which sometimes forms geometrically. But the analysis mentions not a trace of sulphur. Of course our weakness, or impositiveness, lies in that, by anyone to whom it would be agreeable to find sulphur in this thing, sulphur would be found in it—by our own intermediatism there is some sulphur in only a localization or emphasis of something that, unemphasized, is in all things.

So there have, or haven't been found upon this earth things that fell from the sky, or that were left behind by extra–mundane visitors to this earth—

A yarn in the London *Times*, June 22, 1844: that some workmen, quarrying rock, close to the Tweed, about a quarter of a mile below Rutherford Mills, discovered a gold thread embedded in the stone at a depth of eight feet: that a piece of the gold thread had been sent to the office of the *Kelso Chronicle*.

Pretty little thing; not at all frowsy; rather damnable.

London *Times*, Dec. 24, 1851:

That Hiram De Witt, of Springfield, Mass., returning from California, had brought with him a piece of auriferous quartz about the size of a man's fist. It was accidentally dropped—split open—nail in it. There was a cut-iron nail, size of a six-penny nail, slightly corroded. "It was entirely straight and had a perfect head."

Or—California—ages ago when auriferous quartz was forming—super-carpenter, million of miles or so up in the air—drops a nail.

To one not an intermediatist, it would seem incredible that this datum, not only of the damned, or of the journalistic caste of the accursed, could merge away with something else damned only by disregard and backed by what is called "highest scientific authority"—

Communication by Sir David Brewster (*Rept. Brit.*

143

Assoc. 1845–51):

That a nail had been found in a block of stone from Kingoodie Quarry, North Britain. The block in which the nail was found was nine inches thick, but as to what part of the quarry it had come from, there is no evidence—except that it could not have been from the surface. The quarry had been worked about twenty years. It consisted of alternate layers of hard stone and a substance called "till". The point of the nail, quite eaten with rust, projected into some "till", upon the surface of the block of stone. The rest of the nail lay upon the surface of the stone to within an inch of the head—that inch of it was embedded in the stone.

Although its caste is high this is a thing profoundly of the damned—sort of a Brahmin as regarded by a Baptist. Its case was stated fairly; Brewster related all circumstances available to him—but there was no discussion at the meeting of the British Association: no explanation was offered—

Nevertheless the thing can be nullified—

But the nullification that we find is as much against orthodoxy in one respect as it is against our own expression that inclusion in quartz or sandstone indicates antiquity—or there would have to be a revision of prevailing dogmas upon quartz and sandstone and age indicated by them, if the opposing data should be accepted. Of course it may contended by both the orthodox and us heretics that the opposition is only a yarn from a newspaper. By an odd combination we find our two lost souls that have tried to emerge, chucked back to perdition by one blow:

Pop. Sci. News, 1884–41:

That, according to the *Carson Appeal*, there had been found in a mine, quartz crystals that could have had only 15 years in which to form: that, where a mill had been built, sandstone had been found, and when the mill was torn down, that had hardened in 12 years: that in this sandstone was a piece of wood "with a nail in it".

Annals of Scientific Discovery, 1853–71:

That, at the meeting of the British Association, 1853, Sir David Brewster had announced that he had to bring before the meeting an object "of so incredible a nature that nothing short of the strongest evidence was necessary to render the statement at all probable".

A crystal lens had been found in the treasure-house at Nineveh.

In many of the temples and treasure houses of old civilizations upon this earth have been preserved things that have fallen from the sky—or meteorites.

Again we have a Brahmin. This thing is buried alive in the heart of propriety: it is in the British Museum.

Carpenter, in *The Microscope and Its Revelations*, gives two drawings of it. Carpenter argues that it is impossible to accept that optical lenses had ever been made by the ancients.

Never occurred to him—someone a million miles or so up in the air—looking through his telescope—lens drops out.

This does not appeal to Carpenter: he says that this object must have been an ornament.

According to Brewster it was not, an ornament, but "a true optical lens".

In that case, in ruins of an old civilization upon this earth, has been found an accursed thing that was, acceptably not a product of any old civilization indigenous to this earth.

Chapter 10

Early explorers have Florida mixed up with Newfoundland. But the confusion is worse than that still earlier. It arises from simplicity. Very early explorers think that all land westward is one land, India: awareness of other lands as well as India comes as a slow process. I do not now think of things arriving upon this earth from some especial other world. That was my notion when I started to collect our data. Or, as is a commonplace of observation, all intellection begins with the illusion of homogeneity. It's one of Spencer's data: we see homogeneousness in all things distant, or with which we have small acquaintance. Advance from the relatively homogeneous to the relatively heterogeneous is Spencerian Philosophy—like everything else, so-called: not that it was really Spencer's discovery, but was taken from von Baer, who in turn, was continuous with preceding evolutionary speculation. Our own expression is that all things are acting to advance to the homogeneous, or trying to localize Homogeneousness. Homogeneousness is an aspect of the Universal, wherein it is a state that does not merge away into something else. We regard homogeneousness as an aspect of positiveness, but it is our acceptance that infinite frustrations of attempts to positivize manifest themselves in infinite heterogeneity: so that though things try to localize homogeneousness they end up in heterogeneity so great that it amounts to infinite dispersion or indistinguishability.

So all concepts are little attempted positiveness, but soon have to give in to compromise, modification, nullification, merging away into indistinguishability—unless, here and there, in the world's history, there may have been a super-dogmatist, who for only an infinitesimal of time, has been able to hold out against heterogeneity or modification or doubt or "listening to reason", or loss of identity—in which case—instant translation to heaven or the Positive Absolute.

Odd thing about Spencer is that he never recognized that "homogeneity," "integration," and "definiteness" are all words for the same state, or the state that we call "positiveness". What we call his mistake is in that he regarded "homogeneousness" as negative.

I began with a notion of some one other world, from which objects and substances have fallen to this earth; which had, or which, to less degree, has a tutelary interest in this earth; which is now attempting to communicate with this earth—modifying, because of data which will pile up later, into acceptance that some other world is not attempting but has been, for centuries, in communication with a sect, perhaps, or a secret society, or certain esoteric ones of this earth's inhabitants.

I lose a great deal of hypnotic power in not being able to concentrate upon some one other world.

As I have admitted before I'm intelligent, as contrasted with the orthodox. I haven't the aristocratic disregard of a New York curator or an Eskimo medicine-man.

I have to dissipate myself in acceptance of a host of other worlds: size of the moon, some of them: one of them, at least—tremendous thing: we'll take that up later. Vast, amorphous aerial regions, to which such definite words as "worlds" and "planets" seem inapplicable. And artificial constructions that I have called "super-constructions" one of them about the size of Brooklyn, I should say offhand. And one or more of them wheel-shaped things a goodly number of square miles in area.

I think that earlier in this book, before we liberalized into embracing everything that comes along, your indignation, or indigestion would have expressed in the notion that, if this were so, astronomers would have seen these other worlds and regions and vast geometric constructions. You'd have had that notion: you'd have stopped there.

But the attempt to stop is saying "enough" to the insatiable. In cosmic punctuation there are no periods: illusion of periods is incomplete view of colons and semicolons.

We can't stop with the notion that if there were such phenomena, astronomers would have seen them. Because of our experience with suppression and disregard, we suspect, before we go into the subject at all, that astronomers have seen them; that navigators and meteorologists have seen them; that individual scientists have seen them many times—

That it is the System has excluded data of them.

As to the Law of Gravitation, and astronomers' formulas, remember that these formulas worked out in time of Laplace as well as they do now. But there are hundreds of planetary bodies now known that were then not known. So a few hundred worlds more of ours won't make any difference. Laplace knew of about only thirty bodies in this solar system: about six hundred are recognized now—

What are the discoveries of geology and biology to a theologian?

His formulas still work out well as they ever did.

If the Law of Gravitation could be stated as a real utterance, it might be a real resistance to us. But we are told only that gravitation is gravitation. Of course to an intermediatist, nothing can be defined except in terms of itself—but even the orthodox, in what seems to me to be the innate premonitions of realness, not founded upon experience, agree that to define a thing in terms of itself is not real definition. It is said that by gravitation is meant the attraction of all things proportionately to mass and inversely as the square of the distance. Mass would mean inter-attraction holding together final particles, if there were final particles. Then, until final particles be discovered, only one term of this expression survives, or mass is attraction. But distance is only extent of mass, unless one holds out for absolute vacuum among planets, a position against which we could bring a host of data. But there is no possible means of expressing that gravitation is anything other than attraction. So there is nothing to resist us but such a phantom as—that gravitation is the gravitation of all gravitations proportionately to gravitation and inversely as the square of gravitation. In a quasi-existence, nothing more sensible than this can be said upon any so-called subject—perhaps there are higher approximations to ultimate sensibleness.

Nevertheless we seem to have a feeling that with the System against us we have a kind of resistance here. We'd have felt so formerly, at any rate: I think the Dr. Grays and Prof. Hitchcocks have modified our trustfulness toward indistinguishability. As to the perfection of this System that quasi-opposes us and the infallibility of its mathematics—as if there could be real mathematics in a mode of seeming where twice two are not four—we've been told over and over of their vindication in the discovery of

Neptune.

I'm afraid that the course we're taking will turn out like every other development. We began humbly, admitting that we're of the damned.

But our eyebrows—

Just a faint flicker in them, or in one of them, every time we hear of the "triumphal discovery of Neptune"—this "monumental achievement of theoretical astronomy", as the text-books call it.

The whole trouble is that we've looked it up.

The text-books omit this:

That, instead of the orbit of Neptune agreeing with the calculations of Adams and Leverrier, it was so different—that Leverrier said that it was not the planet of his calculations.

Later it was thought best to say no more upon that subject.

The text-books omit this:

That, in 1846, everyone who knew a sine from a cosine was out sining and cosining for a planet beyond Uranus.

Two of them guessed right.

To some minds, even after Leverrier's own rejection of Neptune, the word "guessed" may be objectionable—but, according to Prof. Peirce, of Harvard, the calculations of Adams and Leverrier would have applied quite as well to positions many degrees from the position of Neptune.

Or for Prof. Peirce's demonstration that the discovery of Neptune was only a "happy incident", see *Proc. Amer. Acad. Sciences*, 1–65.

For references, see Lowell's *Evolution of Worlds*.

Or comets: another nebulous resistance to our own notions. As to eclipses, I have notes upon several of them that did not occur upon scheduled time, though with differences only of seconds—and one delightful lost soul, deep-buried, but buried in the ultra-respectable records of the Royal Astronomical Society, upon an eclipse that did not occur at all. That delightful, ultra-sponsored thing of perdition is too good and malicious to be dismissed with passing notice: we'll have him later.

Throughout the history of astronomy, every comet that has come back upon predicted time— not that, essentially, there was anything more abstruse about it than is a prediction that you can make of a postman's periodicities

149

tomorrow—was advertised for all it was worth. It's the way reputations are worked up for fortune-tellers by the faithful. The comets that didn't come back—omitted or explained. Or Encke's comet. It came back slower and slower. But the astronomers explained. Be almost absolutely sure of that they explained. They had it all: worked out and formulated and "proved" why that comet was coming back slower and slower—and there the damn thing began coming faster and faster.

Halley's comet.

Astronomy—"the perfect science, as we astronomers like to call it". (Jackoby.)

It's my own notion that if, in a real existence, an astronomer could not tell one longitude from another, he'd be sent back to this purgatory of ours until he could meet that simple requirement.

Halley was sent to the Cape of Good Hope to determine its longitude. He got it degrees wrong. He gave to Africa's noble Roman promontory a retroussé twist that would take the pride out of any Kaffir.

We hear everlastingly of Halley's comet. It came back—maybe. But, unless we look up in contemporaneous records, we hear nothing of—the Leonids, for instance. By the same methods as those by which Halley's comet was predicted, the Leonids were predicted. November, 1898—no Leonids. It was explained. They had been perturbed. They would appear in November, 1899. November, 1899—November, 1900—no Leonids.

My notion of astronomic accuracy:

Who could not be a prize marksman, if only his hits be recorded?

As to Halley's comet, of 1910—everybody now swears he saw it. He has to perjure himself: otherwise he'd be accused of having no interest in great, inspiring things that he's never given any attention to.

Regard this:

That there never is a moment when there is not some comet in the sky. Virtually there is no year in which several new comets are not discovered, so plentiful are they. Luminous fleas on a vast black dog—in popular impressions, there is no realization of the extent to which this solar system is flea bitten.

If a comet have not the orbit that astronomers have

predicted—perturbed. If—like Halley's comet it be late—even a year late—perturbed. When a train is an hour late, we have small opinion of the predictions of timetables. When a comet's a year late, all we ask is—that it be explained. We hear of the inflation and arrogance of astronomers. My own acceptance is not that they are imposing upon us: that they are requiting us. For many of us priests no longer function to give us seeming rapport with Perfection, Infallibility—the Positive Absolute. Astronomers have stepped forward to fill a vacancy—with quasi-phantomosity—but, in our acceptance, with a higher approximation to substantiality than had the attenuations that preceded them. I should say, myself, that all that we call progress is not so much response to "urge" as it is response to a hiatus—or if you want something to grow somewhere, dig out everything else in its area. So I have to accept that the positive assurances of astronomers are necessary to us, or the blunderings, evasions and disguises of astronomers would never be tolerated: that, given such latitude as they are permitted to take, they could not be very disastrously mistaken. Suppose the comet called Halley's had not appeared—

Early in 1910, a far more important comet than the anaemic luminosity said to be Halley's appeared. It was so brilliant that it was visible in daylight. The astronomers would have been saved anyway. If this other comet did not have the predicted orbit—perturbation. If you're going to Coney Island, and predict there'll be a special kind of a pebble on the beach, I don't see how you can disgrace yourself, if some other pebble will do just as well—because the feeble thing said to have been seen in 1910 was no more in accord with the sensational descriptions given out by astronomers in advance than is a pale pebble with a brick-red boulder.

I predict that next Wednesday, a large Chinaman, in evening clothes, will cross Broadway, at 42nd Street, at 9 P.M. He doesn't, but a tubercular Jap in a sailor's uniform does cross Broadway, at 35th Street, Friday, at noon. Well, a Jap is a perturbed Chinaman, and clothes are clothes.

I remember the terrifying predictions made by the honest and credulous astronomers, who must have been themselves hypnotized, or they could not have hypnotized the rest of us, in 1909. Wills were made. Human life might be

151

swept from this planet. In quasi-existence, which is essentially Hibernian, that would be no reason why wills should not be made. The less excitable of us did expect at least some pretty good fireworks.

I have to admit that it is said that, in New York, a light was seen in the sky.

It was about as terrifying as the scratch of a match on the seat of some breeches half a mile away.

It was not on time.

Though I have heard that a faint nebulosity, which I did not see, myself, though I looked when I was told to look, was seen in the sky, it appeared several days after the time predicted.

A hypnotized host of imbeciles of us; told to look up at the sky: we did—like a lot of pointers hypnotized by a partridge.

The effect.

Almost everybody now swears that he saw Halley's comet, and that it was a glorious spectacle.

An interesting circumstance here is that seemingly we are trying to discredit astronomers because astronomers oppose us—that's not my impression. We shall be in the Brahmin caste of the hell of the Baptists. Almost all our data, in some regiments of this procession, are observations by astronomers, few of them mere amateur astronomers. It is the System that opposes us. It is the System that is suppressing astronomers. I think we pity them in their captivity. Ours is not malice—in a positive sense. It's chivalry—somewhat. Unhappy astronomers looking out from high towers in which they are imprisoned—we appear upon the horizon.

But, as I have said, our data do not relate to some special other world. I mean very much what a savage upon an ocean island might vaguely think of in his speculations—not upon some other land, but complexes of continents and their phenomena: cities, factories in cities, means of communication—

Now all the other savages would know of a few vessels sailing in their regular routes, passing this island in regularized periodicities. The tendency in these minds would be expression of the universal tendency toward positivism—or Completeness—or conviction that these few regularized vessels constituted all. Now I think of some

152

especial savage who suspects otherwise—because he's very backward and unimaginative and insensible to the beautiful ideals of the others: not piously occupied, like the others, in bowing before impressive-looking sticks of wood; dishonestly taking time for his speculations, while the others are patriotically witch-finding. So the other higher and nobler savages know about the few regularized vessels: know when to expect them; have their periodicities all worked out; just about when vessels will pass, or eclipse each other—explaining that all vagaries were due to atmospheric conditions.

They'd come out strong in explaining.

You can't read a book upon savages without noting what resolute explainers they are.

They'd say that all this mechanism was founded upon the mutual attraction of the vessels—deduced from the fall of a monkey from a palm tree—or, if not that, that devils were pushing the vessels—something of the kind.

Storms.

Débris, not from these vessels, cast up by the waves.

Disregarded.

How can one think of something and something else, too.

I'm in the state of mind of a savage who might find upon a shore washed up by the same storm buoyant parts of a piano and a paddle that was carved up by cruder hands than his own: something light and summery from India and a fur overcoat from Russia—or all science though approximating wider and wider is attempt to conceive of India in terms of an ocean island, and of Russia, in terms of India so interpreted. Though I am trying to think of Russia and India in world-wide terms, I cannot think that that, or the universalizing of the local, is cosmic purpose. The higher idealist is the positivist who tries to localize the universal, and is in accord with cosmic purpose: the super-dogmatist of a local savage who can hold out, without a flurry of doubt that a piano washed up on a beach is the trunk of a palm tree that a shark has bitten leaving his teeth in it. So we fear for the soul of Dr. Gray because he did not devote his whole life to that one stand that whether possible or inconceivable thousands of fish had been cast from one bucket.

So unfortunately for myself if salvation be desirable I look out widely but amorphously indefinitely and heterogeneously. If I say I conceive of another world that is now in secret communication with certain esoteric inhabitants of this earth I say I conceive of still other worlds that are trying to establish communication with all the inhabitants of this earth. I fit my notions to the data I find. That is supposed to be the right and logical and scientific thing to do; but it is no way to approximate to form system organization. Then I think I conceive of other worlds and vast structures that pass us by within a few miles without the slightest desire to communicate quite as tramp vessels pass many islands without particularizing one from another. Then I think I have data of a vast construction that has often come to this earth dipped into an ocean, submerged there a while then going away—Why! I'm not absolutely sure. How would an Eskimo explain a vessel sending ashore for coal, which is plentiful upon some Arctic beaches, though of unknown use to the natives, then sailing away, with no interest in the natives?

A great difficulty in trying to understand vast constructions that show no interest in us:

The notion that we must be interesting.

I accept that, though we're usually avoided, probably for moral reasons, sometimes this earth has been visited by explorers. I think that the notion that there have been extra-mundane visitors to China, within what we call the historic period, will be only ordinarily absurd, when we come to that datum.

I accept that some of the other worlds are of conditions very similar to our own. I think of others that are very different—so that visitors from them could not live here—without artificial adaptations.

How some of them could breathe our attenuated air, if they came from a gelatinous atmosphere—

Masks.

The masks that have been found in ancient deposits.

Most of them are of stone, and are said to have been ceremonial regalia of savages—

But the mask that was found in Sullivan County, Missouri, in 1879 (*American antiquarian*, 3–336).

It is made of iron and silver.

Chapter 11

One of the damnedest in our whole saturnalia of the accursed—

Because it is hopeless to try to shake off an excommunication only by saying that we're damned by blacker things than ourselves; and that the damned are those who admit they're of the damned. Inertia and hypnosis are too strong for us. We say that: then we go right on admitting we're of the damned. It is only by being more nearly real that we can sweep away the quasi-things that oppose us. Of course, as a whole, we have considerable amorphousness, but we are thinking now of "individual" acceptances. Wideness is an aspect of Universalness or Realness. If our syntheses disregard fewer data than do opposing syntheses—which are often not syntheses at all, but mere consideration of some one circumstance—less widely synthetic things fade away before us. Harmony is an aspect of the Universal, by which we mean Realness. If we approximate more highly to harmony among the parts of an expression and to all available circumstances of an occurrence, the self—contradictors turn hazy. Solidity is an aspect of realness. We pile them up, and we pile them up, or they pass and pass and pass; things that bulk large as they march by, supporting and solidifying one another—

And still, and for regiments to come, hypnosis and inertia rule us—

One of the damnedest of our data:

In the *Scientific American*, Sept. 10, 1910, Charles F. Holder writes:

"Many years ago, a strange stone resembling a meteorite, fell into the Valley of the Yaqui, Mexico, and the sensational story went from one end to the other of the country that a stone bearing human inscriptions had descended to the earth."

The bewildering observation here is Mr. Holder's assertion that this stone did fall. It seems to me that he must mean that it fell by dislodgement from a mountainside into a valley—but we shall see that it was such a marked stone that very unlikely would it have been unknown to dwellers in a valley, if it had been reposing upon a mountainside above them. It may have been carelessness: intent may have been to say that a sensational story of a strange stone said to have fallen, etc.

This stone was reported by Major Frederick Burnham, of the British Army. Later Major Burnham revisited it, and Mr. Holder accompanied him, their purpose to decipher the incriptions upon it, if possible.

"This stone was a brown, igneous rock, its longest axis about eight feet, and on the eastern face, which had an angle of about forty-five degrees, was the deep-cut inscription."

Mr. Holder says that he recognized familiar Mayan symbols in the inscription. His method was the usual method by which anything can be "identified" as anything else: that is to pick out whatever is agreeable and disregard the rest. He says that he has demonstrated that most of the symbols are Mayan. One of our intermediatist pseudo-principles is that any way of demonstrating anything is just as good a way of demonstrating anything else. By Mr. Holder's method we could demonstrate that we're Mayan—if that should be a source of pride to us. One of the characters upon this stone is a circle within a circle—similar character found by Mr. Holder in a Mayan manuscript. There are two 6's. 6's can be found in Mayan manuscripts. A double scroll. There are dots and there are dashes. Well, then, we, in turn, disregard the circle within a circle and the double scroll and emphasize that 6's occur in this book, and that dots are plentiful, and would be more plentiful if it were customary to use the small "i" for the first personal pronoun—that when it comes to dashes—that's demonstrated: we're Mayan.

I suppose the tendency is to feel that we're sneering at some valuable archaeologic work, and that Mr. Holder did make a veritable identification.

He writes:

"I submitted the photographs to the Field Museum and the Smithsonian and one or two others, and, to my surprise, the reply was that they could make nothing out of it."

Our indefinite acceptance, by preponderance of three or four groups of museum-experts against one person, is that a stone bearing inscriptions unassimilable with any known language upon this earth, is to have fallen from the sky.

Another poor wretch of an outcast belonging here is noted in the *Scientific American*, 48–261: that, of an object, or a meteorite, that fell Feb. 16, 1883, near Brescia, Italy, a false report was circulated that one of the fragments bore the impress of a hand. That's all that is findable by me upon this mere gasp of a thing. Intermediatistically, my acceptance is that, though in the course of human history, there have never been some notable approximations, there never has been a real liar: that he could not survive in intermediateness, where everything merges away or has its pseudo-base in something else—would be instantly translated to the Negative Absolute. So my acceptance is that, though curtly dismissed, there was something to base upon in this report; that there were unusual markings upon this object. Of course that is not to jump to the conclusion that they were cuneiform characters that looked like fingerprints.

Altogether, I think that in some of our past expressions, we must have been very efficient, if the experience of Mr. Symons be typical, so indefinite are we becoming here. Just here we are interested in many things that have been found, especially in the United States, which speak of a civilization, or of many civilizations not indigenous to this earth. One trouble is in trying to decide whether they fell here from the sky, or were left behind by visitors from other worlds. We have a notion there have been disasters aloft, and that coins have dropped here: that inhabitants of this earth found them or saw them fall, and then made coins imitatively: it may be that coins were showered here by something of a tutelary nature that undertook to advance us from the stage of barter to use of a mediam. If coins should be identified as Roman coins, we've had so much experience with "identifications" that we know a phantom when we see one—but, even so, how could Roman coins have got to North America—far in the interior of North America—or buried under the accumulation of centuries of soil—unless they did drop from—wherever the first Romans came from? Ignatius Donnelly, in *Atlantis*, gives a list of objects that have been found in mounds that are supposed to antedate all European influence in America: lathe-made articles, such as traders—from somewhere—would supply to savages—marks of the lathe said to be unmistakable. Said to be: of course we can't accept that anything is unmistakable. In the *Rept. Smithson. Inst.*, 1881–619, there is an account, by Charles C. Jones, of two silver crosses that were found in Georgia. They are skilfully made, highly ornamented crosses, but are not conventional crucifixes: all arms of equal length. Mr. Jones is a good positivist—that De Sota had halted at the "precise" spot where these crosses were found. But the spirit of negativeness that lurks in all things said to be "precise" shows itself in that upon one of these crosses in an inscription that has no meaning in Spanish or any other known, terrestrial language:

"IYNKICIDU," according to Mr. Jones. He thinks that this is a name, and that there is an aboriginal ring to it, though I should say, myself, that he was thinking of the far-distant Incas: that the Spanish donor cut on the cross the name of an Indian to whom it was presented. But we look at the inscription ourselves and see that the letters said to be "C" and "D" are turned the wrong way, and that the letter said to be "K" is not turned the wrong way, but is upside down.

It is difficult to accept that the remarkable, the very extensive, copper mines in the region of Lake Superior were ever the works of American aborigines. Despite the astonishing extent of these mines, nothing has ever been found to indicate that the region was ever inhabited by permanent dwellers— "...not a vestige of a dwelling, a skeleton, or a bone has ever been found". The Indians have no traditions relating to the mines. (*Americ. Antiquarian*, 25–258.) I think that we've had visitors: that they have come here for copper, for instance. As to other relics of them—but we now come upon frequency of a merger that has not so often appeared before:

Fraudulency.

Hair called real hair—then there are wigs. Teeth called real teeth—then there are false teeth. Official money—counter money. It's the bane of psychic research. If there be psychic phenomena, there must be fraudulent psychic phenomena. So desperate is the situation here that Carrington argues that, even if Palladino be caught cheating, that is not to say that all her phenomena are fraudulent. My own version is: that nothing indicates anything, in a positive sense, because, in a positive sense, there is nothing to be indicated. Everything that is called true must merge away indistinguishably into something called false. Both are expressions of the same underlying quasiness, and are continuous. Fraudulent antiquarian relics are very common, but they are not more common than are fraudulent paintings.

W. S. Forest, *Historical Sketches of Norfolk, Virginia:*

That, in September, 1833, when some workmen near Norfolk were boring for water, a coin was drawn up from a depth of about 30 feet. It was about the size of an English shilling, but oval—an oval disc, if not a coin. The figures upon it were distinct, and represented "a warrior or hunter and other characters, apparently of Roman origin".

The means of exclusion would probably be: men digging a hole—no one else looking: one of them drops a coin into the hole: as to where he got a strange coin, remarkable in shape even—that's disregarded. Up comes the coin—expressions of astonishment from the evil one who had dropped it.

However, the antiquarians have missed this coin. I can find no other mention of it.

Another coin. Also a little study in the genesis of a prophet.

In the *American Antiquarian*, 16–313, is copied a story by a correspondent to the *Detroit News*, of a copper coin about the size of a two-cent piece, said to have been found in a Michigan mound. The Editor says merely that he does not endorse the find. Upon this slender basis, he buds out, in the next number of the *Antiquarian*:

"The coin turns out, as we predicted, to be a fraud."

You can imagine the scorn of Elijah, or any of the old more nearly real prophets.

Or all things are tried by the only kind of jurisprudence we have in quasi-existence:

Presumed to be innocent until convicted—but they're guilty.

The Editor's reasoning is as phantom-like as my own, or St. Paul's, or Darwin's. The coin is condemned because it came from the same region from which, a few years before, had come pottery that had been called fraudulent. The pottery had been condemned because it was condemnable.

Scientific American, June 17, 1882:

That a farmer, in Cass Co., Ill., had picked up, on his farm a bronze coin which was sent to Prof. F. F. Hilder, of St.

Louis, who identified it as a coin of Antiochus IV. Inscription said to be in ancient Greek characters: translated as "King Antiochus Epiphanes (illustrious) the Victorius". Sounds quite definite and convincing—but we have some more translations coming.

In the *American Pioneer*, 2–169, are shown two faces of a copper coin, with characters very much like those upon the Grave Creek stone—which, with translations, we'll take up soon. This coin is said to have been found in Connecticut, in 1843.

Records of the Past, 12–182:

That, early in 1913, a coin, said to be a Roman coin was reported as discovered in an Illinois mound. It was sent to Dr. Emerson of the Art Institute, of Chicago. His opinion was that the coin is "of the rare mintage of Domitius Domitianus, Emperor in Egypt." As to its discovery in an Illinois mound, Dr. Emerson disclaims responsibility. But what strikes me here is that a joker should not have been satisfied with an ordinary Roman coin. Where did he get a rare coin, and why was it not missed from some collection? I have looked over numismatic journals enough to accept that the whereabouts of every rare coin in anyone's possession is known to coin-collectors. Seems to me nothing left but to call this another "identification".

Proc. Amer. Phil. Soc., 12–224:

That, in July, 1871, a letter was received from Mr. Jacob W. Moffit, of Chillicothe, Ill., enclosing a photograph of a coin which he said had been brought up, by him, while boring, from a depth of 120 feet.

Of course, by conventional scientific standards such depth has some extraordinary meaning. Paleontologists, geologists, and archaeologists consider themselves reasonable in arguing ancient origin of the far-buried. We only accept: depth is a pseudo-standard with us; one earthquake could bury a coin of recent mintage 120 feet below the surface.

According to a writer in the *Proceedings*, the coin is uniform in thickness, and had never been hammered out by savages—"there are other tokens of the machine shop".

But, according to Prof. Leslie, it is an astrologic amulet. "There are upon it the signs of Pisces and Leo."

Or, with due disregard, you can find signs of your great-

160

grandmother, or of the Crusades, or of the Mayans, upon anything that ever came from Chillicothe or from a five and ten cent store. Anything that looks like a cat and a goldfish looks like Leo and Pisces: but, by due suppressions and distortions there's nothing that can't be made to look like a cat and a goldfish. I fear me we're turning a little irritable here. To be damned by slumbering giants and interesting little harlots and clowns who rank high in their profession is at least supportable to our vanity; but, we find that the anthropologists are of the slums of the divine, or of an archaic kindergarten of intellectuality, and it is very unflattering to find a mess of mouldy infants sitting in judgment upon us.

Prof. Leslie then finds, as arbitrarily as one might find that some joker put the Brooklyn Bridge where it is, that "the piece was placed there as a practical joke, though not by its present owner; and is a modern fabrication, perhaps of the sixteenth century, possibly Hispano–American of French-American origin".

It's sheer, brutal attempt to assimilate a thing that may or may not have fallen from the sky, with phenomena admitted by the anthropologic system: or with the early French or Spanish explorers of Illinois. Though it is ridiculous in a positive sense to give reasons, it is more acceptable to attempt reasons more nearly real than opposing reasons. Of course, in his favour, we note that Prof. Leslie qualifies his notions. But his disregards are that there is nothing either French or Spanish about this coin. A legend upon it is said to be "somewhere between Arabic and Phoenician, without being either". Prof. Winchell (*Sparks from a Geologist's Hammer*, p.170) says of the crude designs upon this coin, which was in his possession—scrawls of an animal and of a warrior, or of a cat and a goldfish, whichever be convenient—that they had been neither stamped nor engraved, but "looked as if etched with an acid". That is a method unknown in numismatics of this earth. As to the crudity of design upon this coin, and something else—that, though the "warrior" may be, by due disregards, either a cat or a goldfish, we have to note that his headdress is typical of the American Indian—could be explained, of course, but for fear that we might be instantly translated to the Positive Absolute, which may not be absolutely desirable, we

161

prefer to have some flaws or negativeness in our own expressions.

Data of more than the thrice-accursed:

Tablets of stone, with the ten commandments engraved upon them, in Hebrew, said to have been found in mounds in the United States;

Masonic emblems said to have been found in mounds in the United States.

We're upon the borderline of our acceptances, and we're amorphous in the uncertainties and mergings of our outline. Conventionally, or, with no real reason for so doing, we exclude these things, and then, as grossly and arbitrarily and irrationally—though our attempt is always to approximate away from these negative states—as ever a Kepler, Newton, or Darwin made his selections, without which he could not have seemed to be, at all, because every one of them is now seen to be an illusion, we accept that other lettered things have been found in mounds in the United States. Of course we do what we can to make the selection seem not gross and arbitrary and irrational. Then, if we accept that inscribed things of ancient origin have been found in the United States; that cannot be attributed to any race indigenous to the western hemisphere; that are not in any language ever heard of in the eastern hemisphere—there's nothing to it but to turn non-Euclidian and try to conceive of a third "hemisphere", or to accept that there has been intercourse between the western hemisphere and some other world.

But there is a peculiarity to these inscribed objects. They remind me of the records left, by Sir John Franklin, in the Arctic; but, also, of attempts made by relief expeditions to communicate with the Franklin expedition. The lost explorers cached their records—or concealed them conspicuously in mounds. The relief expeditions sent up balloons, from which messages were dropped broadcast. Our data are of things that have been cached, and of things that seem to have been dropped—

Or a Lost Expedition from—Somewhere.

Explorers from somewhere, and their inability to return—then, a long, sentimental, persistent attempt, in the spirit of our own Arctic relief: expeditions—at least to establish communication—

What if it may have succeeded?

We think of India—the millions of natives who are ruled by a small band of esoterics—only because they receive support and direction from—somewhere else—or from England.

In 1838, Mr. A. B. Tomlinson, owner of the great mound at Grave Creek, West Virginia, excavated the mound. He said that, in the presence of witnesses he had found a small, flat, oval stone—or disc—upon which were engraved alphabetic characters.

Col. Whittelsey, an expert in these matters, says that the stone is now "universally regarded by archaeologists as a fraud": that, in his opinion, Mr. Tomlinson had been imposed upon.

Avebury, *Prehistoric Times*, p. 271:

"I mention it because it has been the subject of much discussion, but it is now generally admitted to be a fraud. It is inscribed with Hebrew characters, but the forger has copied the modern instead of the ancient form of the letters."

As I have said, we're as irritable here, under the oppressions of the anthropologists as ever were slaves in the south toward superiorities from "poor white trash". When we finally reverse our relative positions we shall give lowest place to the anthropologists. A Dr. Gray does at least look at a fish before he conceives of a miraculous origin for it. We shall have to submerge Lord Avebury far below him—if we accept that the stone from Grave Creek is generally regarded as a fraud by eminent authorities who did not know it from some other object—or, in general, that so decided an opinion must be the product of either deliberate disregard or ignorance or fatigue. The stone belongs to a class of phenomena that is repulsive to the System. It will not assimilate with the System. Let such an object be heard of by such a systematist as Avebury, and the mere mention of it is as nearly certainly the stimulus to a conventional reaction as is a charged body to an electroscope or a glass of beer to a prohibitionist. It is of the ideals of Science to know one subject from another before expressing an opinion upon a thing, but that is not the spirit of universal mechanics.

A thing. It is attractive or repulsive. Its conventional reaction follows.

Because it is not the stone from Grave Creek that is in

Hebrew characters, either ancient or modern: it is a stone from Newark, Ohio, of which the story is told that a forger made this mistake of using modern instead of ancient Hebrew characters. We shall see that the inscription upon the Grave Creek stone is not in Hebrew.

Or all things are presumed to be innocent, but are supposed to be guilty—unless they assimilate.

Col. Whittelsey (*Western Reserve Historical Tracts, No. 33*) says that the Grave Creek stone was considered a fraud by Wilson, Squires, and Davis. Then he comes to the Congress of Archaeologists at Nancy, France, 1875. It is hard for Col. Whittelsey to admit that, at this meeting, which sounds important, the stone was endorsed. He reminds us of Mr. Symons, and "the man" who "considered" that he saw something. Col. Whittelsey's somewhat tortuous expression is that the finder of the stone "so imposed his views" upon the congress that it pronounced the stone genuine.

Also the stone was examined by Schoolcraft. He gave his opinion for genuineness.

Or there's only one process, and "see-saw" is one of its aspects. Three or four fat experts on the side against us. We find four or five plump ones on our side. Or all that we call logic and reasoning ends up as sheer preponderance of avoirdupois.

Then several philologists came out in favour of genuineness. Some of them translated the inscription. Of course, as we have said, it is our method—or the method of orthodoxy—way in which all conclusions are reached—to have some way in which all conclusions are reacted—to have some awfully eminent, or preponderantly plump, authorities with us whenever we can—in this case, however, we feel just a little apprehensive in being caught in such excellently obese, but somewhat negativized, company:

Translation by M. Jombard:

"Thy orders are laws: though shinest in impetuous élan and rapid chamois."

M. Maurice Schwab:

"The chief of Emigration who reached these places (or this island) has fixed these characters forever."

M. Oppert:

"The grave of one who was assassinated here. May God,

164

to revenge him, strike his murderer, cutting off the hand of his existence."

I like the first one best. I have such a vivid impression from it of someone polishing up brass or something, and in an awful hurry. Of course the third is more dramatic—still they're all very good. They are perturbations of one another, I suppose.

In Tract 44, Col. Whittelsey returns to the subject. He gives the conclusion of Major De Helward, at the Congress of Luxembourg, 1877:

"If Prof. Read and myself are right in the conclusion that the figures are neither of the Runic, Phoenician, Canaanite, Hebrew, Lybian, Celtic, or any other alphabet-language, its importance has been greatly over-rated."

Obvious to a child; obvious to any mentality not helplessly subjected to a system:

That just therein lies the importance of this object.

It is said that an ideal of science is to find out the new—but unless a thing be of the old, it is "unimportant".

"It is not worth while." (Hovey.)

Then the inscribed axe, or wedge, which, according to Dr. John C. Evans, in a communication to the American Ethnological Society, was ploughed up, near Pemberton, N.J., 1859. The characters upon this axe or wedge, are strikingly similar to the characters on the Grave Creek stone. Also, with a little disregard here and a little more there, they look like tracks in the snow by someone who's been out celebrating, or like your handwriting, or mine, when we think there's a certain distinction in illegibility. Method of disregard: anything's anything.

Dr. Abbott describes this object in the *Report of the Smithsonian Institution,* 1875–260.

He says he has no faith in it.

All progress is from the outrageous to the commonplace. Or quasi-existence proceeds from rape to the crooning of lullabies. It's been interesting to me to go over various long-established periodicals and note controversies between attempting positivists and then intermediatistic intentions—the alarms of Science; her attempts to preserve that which is dearer than life itself—submission—then a fidelity like Mrs. Micawber's. So many of these ruffians, or wandering comedians that were hated, or scorned, pitied, embraced, conventionalized. There's not a notion in this book that has a more frightful, or ridiculous, mien than had the notion of human footprints in rocks, when that now respectabilized ruffian, or clown, was first heard from. It seems bewildering to one whose interests are not scientific that such rows should be raised over such trifles: but the feeling of a systematist toward such an intruder is just about what anyone's would be if a tramp from the street should come in, sit at one's dinner table, and say that he belonged there. We know what hypnosis can do: let him insist with all his might that he does belong there, and one begins to suspect that he may be right; that he may have higher perceptions of what's right. The prohibitionists had this worked out very skilfully.

So the row that was raised over the stone from Grave Creek—but time and cumulativeness, and the very factor we make so much of—or the power of massed data. There were other reports of inscribed stones, and then, half a century later, some mounds—or caches, as we call them—were opened by the Rev. Mr. Gass, near the city of Davenport. (*American Antiquarian*, 15–73.) Several stone tablets were found. Upon one of them, the letters "TFTOWNS" may easily be made out. In this instance we hear nothing of fraudulency—time, cumulativeness, the power of massed data. The attempt to assimilate this datum is:

That the tablet was probably of Mormon origin.

Why?

Because, at Mendon, Ill., was found a brass plate, upon which were similar characters.

Why that?

Because that was found "near the house once occupied by a Mormon".

In a real existence, a real meteorologist, suspecting that cinders had come from a fire engine—would have asked a fireman.

Tablets of Davenport—there's not a record findable that it ever occurred to any antiquarian—to ask a Mormon.

Other tablets were found. Upon one of them are two "F's" and two "8's". Also a large tablet, twelve inches by eight to ten inches "with Roman numerals and Arabic". It is said that the figure "8" occurs three times, and the figure or letter "O" seven times. "With these familiar characters are others that resemble ancient alphabets either

Phoenecian or Hebrew."

It may be that the discovery of Australia, for instance, will turn out to be less important than the discovery and the meaning of these tablets—But where will you read of them in anything subsequently published; what antiquarian has ever since tried to understand them, and their presence, and indications of antiquity, in a land that we're told was inhabited only by unlettered savages?

These things that are exhumed only to be buried in some other way.

Another tablet was found, at Davenport, by Mr. Charles Harrison, president of the American Antiquarian Society. "... 8 and other hieroglyphics are upon this tablet". This time, also, fraud is not mentioned. My own notion is that it is very unsportsmanlike ever to mention fraud. Accept anything. Then explain it your way. Anything that assimilates with one explanation, must have assimilable relations, to some degree, with all other explanations, if all explanations are somewhere continuous. Mormons are lugged in again, but the attempt is faint and helpless—"because general circumstances make it difficult to explain the presence of these tablets".

Altogether our phantom resistance is mere attribution to the Mormons, without the slightest attempt to find base for the attribution. We think of messages that were showered upon this earth, of messages that were cached in mounds upon this earth. The similarity to the Franklin situation is striking. Conceivable centuries from now, objects dropped from relief-expedition balloons may be found in the Arctic, and conceivably there are still undiscovered caches left by Franklin, in the hope that relief expeditions would find them. It would be as incongruous to attribute these things to the Eskimos as to attribute tablets and lettered stones to the aborigines of America. Some time I shall take up an expression that the queer-shaped mounds upon this earth were built by explorers from Somewhere, unable to get back, designed to attract attention from some other world, and that a vast sword-shaped mound has been discovered upon the moon—Just now we think of lettered things and their two possible significances.

A bizarre little soul, rescued from one of the morgues of the *American Journal of Science:*

An account, sent by a correspondent, to Prof. Silliman, of

something that was found in a block of marble, taken November, 1829, from a quarry, near Philadelphia (*Am. J. Sci.*, 1–19–361). The block was cut into slabs. By this process, it is said, was exposed an indentation in the stone, about one and a half inches by five-eighths of an inch. A geometric indentation: in it were two definite-looking raised letters, like "I U": only difference is that the corners of the "U" are not rounded, but are right angles. We are told that this block of stone came from a depth of 70 or 80 feet—or that, if acceptable, this lettering was done long, long ago. To some persons, not sated with the commonness of the incredible that has to be accepted, it may seem grotesque to think that an indentation in sand could have tons of other sand piled upon it and hardening into stone, without being pressed out—but the famous Nicaraguan footprints were found in a quarry under eleven strata of solid rock. There was no discussion of this datum. We only take it out for an airing.

As to lettered stones that may once upon a time have been showered upon Europe, if we cannot accept that the stones were inscribed by indigenous inhabitants of Europe, many have been found in caves—whence they were carried as curiosities by prehistoric men, or as ornaments, I suppose. About the size and shape of the Grave Creek stone, or disc: "flat and oval and about two inches wide". (Sollas.) Characters painted upon them: found first by M. Piette, in cave of Mas d'Azil, Ariège. According to Sollas, they are marked in various directions with red and black lines. "But on not a few of them, more complex characters occur, which in a few instances simulate some of the capital letters of the Roman alphabet." In one instance the letters "FEI" accompanied by no other markings to modify them, are as plain as they could be. According to Sollas (*Ancient Hunters*, p. 95) M. Cartailhac has confirmed the observations of Piette, and M. Boule has found additional examples. "They offer one of the darkest problems of prehistoric times." (Sollas.)

As to caches in general, I should say that they are made with two purposes: to proclaim and to conceal; or that caches documents are hidden, or covered over, in conspicuous structures; at least, so are designed the cairns in the Arctic.

Trans. N.Y. Acad. of Sciences, 11–27:

That Mr. J. H. Hooper, Bradley Co., Tenn., having come upon a curious stone, in some woods upon his farm, investigated. He dug. He unearthed a long wall. Upon this wall were inscribed many alphabetic characters. "872 characters have been examined, many of them duplicates, and a few imitations of animal forms, the moon, and other objects. Accidental imitations of oriental alphabets are numerous."

The part that seems significant:

That these letters had been hidden under a layer of cement.

And still, in our own heterogeneity, or unwillingness, or inability, to concentrate upon single concepts, we shall—or we shan't—accept that, though there may have been a Lost Colony of Lost Expedition from Somewhere, upon this earth, and extra-mundane visitors who could never get back, there have been other extra-mundane visitors, who have gone away again—altogether quite in analogy with the Franklin Expedition and Peary's flittings in the Arctic—

And a wreck that occurred to one group of them—

And the loot that was lost overboard—

The Chinese seals of Ireland.

Not the things with the big, wistful eyes that lie on ice, and that are taught to balance objects on their noses—but inscribed stamps, with which to make impressions.

Proc. Roy. Irish Acad., 1–381:

A paper was read by Mr. J. Huband Smith, descriptive of about a dozen Chinese seals that had been found in Ireland. They are all alike: each a cube with an animal seated upon it. "It is said that the inscriptions upon them are of a very ancient class of Chinese characters."

The three points that have made a leper and an outcast of this datum—but only in the sense of disregard, because nowhere that I know of is it questioned:

Agreement among archaeologists that there were no relations in the remote past, between China and Ireland:

That no other objects, from ancient China—virtually, I suppose—have ever been found in Ireland;

The great distances at which these seals have been found apart.

After Mr. Smith's investigations—if he did investigate, or do more than record—many more Chinese seals were found in Ireland, and, with one exception, only in Ireland. In 1852, about 60 had been found. Of all archaeological finds in Ireland, "none is enveloped in greater mystery". (*Chambers' Journal*, 16–364.) According to the writer in *Chambers' Journal*, one of these seals was found in a curiosity shop in London. When questioned, the shopkeeper said that it had come from Ireland.

In this instance, if you don't take instinctively to our expression, there is no orthodox explanation for your preference. It is the astonishing scattering of them, over field and forest, that has hushed the explainers. In the *Proceedings of the Royal Irish Academy*, 10–171, Dr. Frazer says that they "appear to have been sown broadcast over the country in some strange way that I cannot offer solution of".

The struggle for expression of a notion that did not belong to Dr. Frazer's era:

"The invariable story of their find is what we might expect if they had been accidentally dropped ..."

Three were found in Tipperary; six in Cork; three in Down; four in Waterford; all the rest—one or two to a county.

But one of these Chinese seals was found in the bed of the River Boyne, near Clonard, Meath, when workmen were raising gravel.

That one, at least, had been dropped there.

Chapter 12

Astronomy.

And a watchman looking at half a dozen lanterns, where a street's been torn up.

There are gas lights and kerosene lamps and electric lights in the neighbourhood: matches flaring, fires in stoves, bonfires, house afire somewhere; lights of automobiles, illuminated signs—

The watchman and his little system.

Ethics.

And some young ladies and the dear old professor of a very "select" seminary.

Drugs and divorce and rape: venereal diseases, drunkenness, murder—

Excluded.

The prim and the precise, or the exact, the homogeneous, the single, the puritanic, the mathematical, the pure, the perfect. We can have illusion of this state—but only by disregarding its infinite denials. It's a drop of milk afloat in acid that's eating it. The positive swamped by the negative. So it is in intermediateness, where only to "be" positive is to generate corresponding and, perhaps, equal negativeness. In our acceptance, it is, in quasi-existence, premonitory, or prenatal, or pre-awakening consciousness of a real existence.

But this consciousness of realness is the greatest resistance to efforts to realize or to become real—because it is feeling that realness has been attained. Our antagonism is not to Science, but to the attitude of the sciences that they have finally realized; or to belief, instead of acceptance; to the insufficiency, which, as we have seen over and over, amounts to paltriness and puerility of scientific dogmas and standards. Or, if several persons start out to Chicago, and get to Buffalo, and one be under the delusion that Buffalo is Chicago, that one will be a resistance to the progress of the others.

So astronomy and its seemingly exact, little system—

But data we shall have of round worlds and spindle-shaped worlds, and worlds shaped like a wheel; worlds like titanic pruning hooks; worlds linked together by streaming filaments; solitary worlds, and worlds in hordes: tremendous worlds and tiny worlds: some of them made of material like the material of this earth; and worlds that are geometric super-constructions made of iron and steel—

Or not only fall from the sky of ashes and cinders and coke and charcoal and oily substances that suggest fuel—but the masses of iron that have fallen upon this earth.

Wrecks and flotsam and fragments of vast iron constructions—

Or steel. Sooner or later we shall have to take up an expression that fragments of steel have fallen from the sky. If fragments not of iron, but of steel have fallen upon this earth—

But what would a deep-sea fish learn even if a steel plate of a wrecked vessel above him should drop and bump him on the nose?

Our submergence in a sea of conventionality of almost impenetrable density.

Sometimes I'm a savage who has found something on the beach of his island. Sometimes I'm a deep-sea fish with a sore nose.

The greatest of mysteries:

Why don't they ever come here, or send here, openly?

Of course there's nothing to that mystery if we don't take so seriously the notion—that we must be interesting. It's probably for moral reasons that they stay away—but even so, there must be some degraded ones among them.

Or physical reasons:

When we can specially take up that subject, one of our leading ideas, or credulities, will be that near approach by another world to this world would be catastrophic: that navigable worlds would avoid proximity; that others that have survived have organized into protective remoteness, or orbits which approximate to regularity, though by no means to the degree of popular supposition.

But the persistence of the notion that we must be interesting. Bugs and germs and things like that: they're interesting to us: some of them are too interesting.

Dangers of near approach—nevertheless our own ships that dare not venture close to a rocky shore can send rowboats ashore—

172

Why not diplomatic relations established between the United States and Cyclorea—which, in our advanced astronomy, is the name of a remarkable wheel-shaped world or super-construction? Why not missionaries sent here openly to convert us from our barbarous prohibitions and other taboos, and to prepare the way for a good trade in the ultra-bibles and super-whiskies; fortunes made in selling us cast-off super-fineries, which we'd take to like an African chief to someone's old silk hat from New York or London?

The answer that occurs to me is so simple that it seems immediately acceptable, if we accept that the obvious is the solution of all problems, or if most of our perplexities consist in laboriously and painfully conceiving of the unanswerable, and then looking for answers—using such words as "obvious" and "solution" conventionally—

Or:

Would we, if we could, educate and sophisticate pigs, geese, cattle?

Would it be wise to establish diplomatic relation with the hen that now functions, satisfied with mere sense of achievement by way of compensation?

I think we're property.

I should say we belong to something:

That once upon a time, this earth was No-man's Land, that other worlds explored and colonized here, and fought among themselves for possession, but now it's owned by something:

That something owns this earth—all others warned off.

Nothing in our times—perhaps—because I am thinking of certain notes I have—has ever appeared upon this earth, from somewhere else, so openly as Columbus landed upon San Salvador, or as Hudson sailed up his river. But as to surreptitious visits to this earth, in recent times, or as to emissaries, perhaps, from other worlds, or voyagers who have shown every indication of intent to evade and avoid, we shall have data as convincing as our data of oil or coal-burning aerial super-constructions.

But, in this vast subject, I shall have to do considerable neglecting or disregarding, myself. I don't see how I can, in this book, take up at all the subject of possible use of humanity to some other mode of existence, or the flattering notion that we can possibly be worth something.

173

Pigs, geese, and cattle.

First find out that they are owned.

Then find out the whyness of it.

I suspect that, after all, we're useful—that among contesting claimants, adjustment has occurred, or that something now has a legal right to us, by force, or by having paid out analogues of beads for us to former, more primitive, owners of us—all others warned off—that all this has been known, perhaps for ages, to certain ones upon this earth, a cult or order, members of which function like bell-wethers to the rest of us, or as superior slaves or overseers, directing us in accordance with instructions received—from Somewhere else—in our mysterious usefulness.

But I accept that, in the past, before proprietorship was established, inhabitants of a host of other worlds have—dropped here, hopped here, wafted, sailed, flown, motored—walked here for all I know—been pulled here been pushed; have come singly, have come in enormous numbers; have visited occasionally, have visited periodically for hunting, trading, replenishing harems, mining: have been unable to stay here, have established colonies here, have been lost here; far-advanced peoples, or things, and primitive peoples or whatever they were: white ones, black ones, yellow ones—

I have a very convincing datum that the ancient Britons were blue ones.

Of course we are told by conventional anthropologists that they only painted themselves blue, but in our own advanced anthropology, they were veritable blue ones—

Annals of Philosophy, 14-51:

Note of a blue child born in England.

That's atavism.

Giants and fairies. We accept them, of course. Or, if we pride ourselves upon being awfully far-advanced, I don't know how to sustain our conceit except by very largely going far back. Science of today—the superstition of tomorrow. Science of tomorrow—the superstition of today.

Notice of a stone axe, 17 inches long: 9 inches across broad end. (*Proc. Soc. of Ants. of Scotland*, 1–9–184.)

Amer. Antiquarian, 18–60:

Copper axe from an Ohio mound: 22 inches long; weight 38 pounds.

Amer. Anthropologist, n.s., 8–229:

Stone axe found at Birchwood, Wisconsin—exhibited in the collection of the Missouri Historical Society—found with "the pointed end embedded in the soil"—for all I know, may have dropped there—28 inches long, 14 wide, 11 thick—weight 300 pounds.

Or the footprints, in the sandstone, near Carson, Nevada—each print 18–20 inches long. (*Amer. Jour. Sci.*, 3-26-139.)

These footprints are very clear and well-defined: reproduction of them in the *Journal*—but they assimilate with the System, like sour apples to other systems: so Prof. Marsh, a loyal and unscrupulous systematist, argues:

"The size of these footprints and specially the width between the right and left series, are strong evidence that they were not made by men, as has been so generally supposed."

So these excluders. Stranglers of Minerva. Desperadoes of disregard. Above all, or below all, the anthropologists. I'm inspired with a new insult—someone offends me: I wish to express almost absolute contempt for him—he's a systematic anthropologist. Simply to read something of this kind is not so impressive as to see for one's self: if anyone will take the trouble to look up these footprints, as pictured in the *Journal*, he will either agree with Prof. Marsh or feel that to deny them is to indicate a mind as profoundly enslaved by a system as was ever the humble intellect of a mediaeval monk. The reasoning of this representative phantom of the chosen, or of the spectral appearances who sit in judgment, or condemnation, upon us of the more nearly real:

That there never were giants upon this earth, because gigantic footprints are more gigantic than prints made by men who are not giants.

We think of giants as occasional visitors to this earth. Of course—Stonehenge, for instance. It may be that, as time goes on, we shall have to admit that there are remains of many tremendous habitations of giants upon this earth, and that their appearances here were more than casual—but their bones—or the absence of their bones—

Except—that, no matter how cheerful and unsuspicious my disposition may be, when I go to the American Museum of Natural History, dark cynicisms arise the moment I come to the fossils—or old bones that have been found upon this earth—gigantic things—that have been reconstructed into terrifying but "proper" dinosaurs—but my uncheerfulness—

The dodo did it.

On one of the floors below the fossils, they have a reconstructed dodo. It's frankly a fiction: it's labelled as such—but it's been reconstructed so cleverly and so convincingly—

Fairies.

"Fairy crosses."

Harper's Weekly, 50–715:

That, near the point where the Blue Ridge and the Allegheny Mountains unite, north of Patrick County, Virginia, many little stone crosses have been found.

A race of tiny beings.

They crucified cockroaches.

Exquisite beings—but the cruelty of the exquisite. In their diminutive way they were human beings. They crucified.

The "fairy crosses", we are told in *Harper's Weekly*, range in weight from one-quarter of an ounce to an ounce: but it is said, in the *Scientific American*, 79–395, that some of them are no larger than the head of a pin.

They have been found in two other states, but all in Virginia are strictly localized on and along Bull Mountain.

We are reminded of the Chinese seals in Ireland.

I suppose they fell there.

Some are Roman crosses, some St. Andrew's, some Maltese. This time we are spared contact with the anthropologists and have geologists instead, but I am afraid that the relief to our finer, or more nearly real, sensibilities will not be very great. The geologists were called upon to explain the "fairy crosses". Their response was the usual scientific tropism—"Geologists say that they are crystals." The writer in *Harper's Weekly* points out that this "hold up", or this anaesthetic, if theoretic science be little but attempt to assuage pangs of the unexplained, fails to account for the localized distributions of these objects—which makes me think of both aggregation and separation at the bottom of the sea, if from a wrecked ship, similar objects should fall in large numbers but at different times.

But some are Roman crosses, some St. Andrew's, some

Maltese.

Conceivably there might be a mineral that would have a diversity of geometric forms, at the same time restricted to some expression of the cross, because snowflakes, for instance, have diversity but restriction to the hexagon, but the guilty geologists, cold-blooded as astronomers and chemists and all the other deep-sea fishes—though less profoundly of the pseudo-saved than the wretched anthropologists—disregarded the very datum—that it was wise to disregard:

That the "fairy crosses" are not all made of the same material.

It's the same old disregard, or it's the same old psychotropism, or process of assimilation. Crystals are geometric forms. Crystals are included in the System. So then "fairy crosses" are crystals. But that different minerals should, in a few different regions, be inspired to turn into less nearly real than our own acceptances.

We now come to some "cursed" little things that are of the "lost", but the "salvation" of which scientific missionaries have done their damnedest.

"Pygmy flints."

They can't very well be denied.

They're lost and well known.

"Pygmy flints" are tiny, prehistoric implements. Some of them are a quarter of an inch in size, England, India, France, South Africa—they've been found in many parts of the world—whether showered there or not. They belong high up in the froth of the accursed: they are not denied, and they have not been disregarded; there is an abundant literature upon this subject. One attempt to rationalize them, or assimilate them, or take them into the scientific fold, has been the notion that they were toys of prehistoric children.

It sounds reasonable. But, of course, by the reasonable we mean that for which the equally reasonable, but opposing, has not been found out—except that we modify that by saying that, though nothing's finally reasonable, some phenomena have higher approximations to Reasonableness than have others. Against the notion of toys, the higher approximation is that where "pygmy flints" are found, all flints are pygmies—at least so in India, where, when larger implements have been found in the same place, there are separations by strata. (Wilson.)

The datum that, just at present, leads me to accept that these flints were made by beings about the size of pickles, is a point brought out by Prof. Wilson (*Rept. National Museum*, 1892–455):

Not only that the flints are tiny but that the chipping upon them is "minute".

Struggle for expression, in the mind of a nineteenth-century-ite, of an idea that did not belong to his era:

In *Science Gossip*, 1896–36, R. A. Galty says:

"So fine is the chipping that to see the workmanship a magnifying glass is necessary."

I think that would be absolutely convincing, if there were anything—absolutely anything—either that tiny beings, from pickle to cucumber-stature, made these things, or that ordinary savages made them under magnifying glasses.

The idea that we are now going to develop, or perpetrate, is rather intensely of the accursed, or the advanced. It's a lost soul, I admit or boast—but it fits in. Or, as conventional as ever, our own method is the scientific method of assimilating. It assimilates, if we think of the inhabitants of Elvera—

By the way, I forget to tell the name of the giant's world: Monstrator.

Spindle-shaped world—about 100,000 miles along its major axis—more details to be published later.

But our coming inspiration fits in, if we think of the inhabitants of Elvera as having only visited here: having, in hordes as dense as clouds of bats, come here, upon hunting excursions—for mice, I should say: for bees, very likely—or most likely of all, or inevitably, to convert the heathen here—horrified with anyone who would gorge himself with more than a bean at a time; fearful for the souls of beings who would guzzle more than a dewdrop at a time—hordes of tiny missionaries, determined that right should prevail, determining right by their own minute-nesses.

They must have been missionaries.

Only to be is motion to convert or assimilate something else.

The idea now is that tiny creatures coming here from their own little world, which may be Eros, though I call it Elvera, would flit from the exquisite to the enormous—gulp of a fair-sized terrestrial animal—half a dozen of them gone and soon digested. One falls into a brook—torn away in a mighty torrent—

Or never anything but conventional, we adopt from Darwin:

"The geological records are incomplete."

Their flints would survive, but, as to their fragile bodies—one might as well search for prehistoric frost-traceries. A little whirlwind—Elverean carried away a hundred yards—body never found by his companions. They'd mourn for the departed. Conventional emotion to have: they'd mourn. There'd have to be a funeral: there's no getting away from funerals. So I adopt an explanation that I take from the anthropologists: burial in effigy. Perhaps the Elvereans would not come to this earth until many years later—another distressing occurrence—one little mausoleum for all burials in effigy.

London *Times*, July 20, 1836:

That, early in July, 1836, some boys were searching for rabbits' burrows in the rocky formation, near Edinburgh, known as Arthur's Seat. In the side of a cliff, they came upon some thin sheets of slate, which they pulled out.

Little cave.

Seventeen tiny coffins.

Three or four inches long.

In the coffins were miniature wooden figures. They were dressed differently both in style and material. There were two tiers of eight coffins each, and a third tier begun, with one coffin.

The extraordinary datum, which has especially made mystery here:

That the coffins had been deposited singly, in the little cave, and at intervals of many years. On the first tier, the coffins were quite decayed, and the wrappings had mouldered away. In the second tier, the effects of age had not advanced so far. And the top coffin was quite recent-looking.

In the *Proceedings of the Society of Antiquarians of Scotland*, 3-12-460, there is a full account of this find. Three of the coffins and three of the figures are pictured.

So Elvera with its downy forests and its microscopic oyster shells—and if the Elvereans be not very far-advanced, they take baths—with sponges the size of pin heads—

Or that catastrophes have occurred: that fragments of Elvera have fallen to this earth:

In *Popular Science*, 20–83, Francis Bingham, writing of the corals and sponges and shells and crinoids that Dr. Hahn had asserted that he had found in meteorites, says, judging by the photographs of them, that their "notable peculiarity" is their "extreme smallness". The corals, for instance, are about one-twentieth the size of terrestrial corals. "They represent a veritable pygmy animal world", says Bingham.

The inhabitants of Monstrator and Elvera were primitives, I think, at the time of their occasional visits to this earth—though, of course, in quasi-evidence, anything that we semi-phantoms call evidence of anything may be just as good evidence of anything else. Logicians and detectives and jurymen and suspicious wives and members of the Royal Astronomic Society recognize this indeterminateness, but have the delusion that in the method of agreement there is final, or real evidence. The method is good enough for an "existence" that is only semi-real, but also it is the method of reasoning by which witches were burned, and by which ghosts have been feared. I'd not like to be so unadvanced as to deny witches and ghosts like those of popular supposition. But stories of them have been supported by astonishing fabrications of details and of different accounts in agreement.

So, if a giant left impressions of his bare feet in the ground, that is not to say that he was a primitive—bulk of culture out taking the Kneipp cure. So, if Stonehenge is a large, but only roughly geometric construction, the inattention to details by its builders—signifies anything you please —ambitious dwarfs or giants—if giants, that they were little more than cave men, or that they were post-impressionist architects from a very far-advanced civilization.

If there are other worlds, there are tutelary worlds—or that Kepler, for instance, could not have been absolutely wrong: that his notion of an angel assigned to push along and guide each planet may not be very acceptable, but that, abstractedly, or in the notion of a tutelary relation, we may find acceptance.

Only to be is to be tutelary.

Our general expression:

That "everything" in Intermediateness is not a thing, but is an endeavour to become something—by breaking away from its continuity, or merging away, with all other phenomena—is an attempt to break away from the very essence of a relative existence and become absolute—if it have not surrendered to, or become part of, some higher attempt:

That to this process there are two aspects:

Attraction, or the spirit of everything to assimilate all other things—if it have not given in and subordinated to—or have not been assimilated by—some higher attempted system, unity, organization, entity, harmony, equilibrium—

And repulsion, or the attempt of everything to exclude or disregard the unassimilable.

Universality conceivable:

A tree. It is doing all it can to assimilate substances of the soil and substances of the air, and sunshine, too, into tree-substance: obversely it is rejecting or excluding or disregarding that which it cannot assimilate.

Cow grazing, pig rooting, tiger stalking: planets trying, or acting, to capture comets; rag pickers and the Christian religion, and a cat down headfirst in a garbage can; nations fighting for more territory, sciences correlating the data they can, trust magnates organizing, chorus girl out for a little late supper—all of them stopped somewhere by the unassimilable. Chorus girl and the broiled lobster. If she eats not shell and all she represents universal failure to positivize: her ensuing disorders will translate her to the Negative Absolute.

Or Science and some of our cursed hard-shelled data.

One speaks of the tutelarian as if it were something distinct in itself. So one speaks of a tree, a saint, a barrel of pork, the Rocky Mountains. One speaks of missionaries, as if they were positively different, or had identity of their own, or were a species by themselves. To the Intermediatist, everything that seems to have identity is only attempted identity, and every species is continuous with all other species, or that which is called the specific is only emphasis upon some aspect of the general. If there are cats, they're only emphasis upon universal felinity. There is nothing that does not partake of that of which the missionary, or the tutelary, is the special. Every conversation is a conflict of missionaries, each trying to convert the other, to assimilate, or to make the other similar to himself. If no progress be made, mutual repulsion will follow.

If other worlds have ever in the past had relations with this earth, they were attempted positivizations: to extend themselves, by colonies, upon this earth; to convert, or assimilate, indigenous inhabitants of this earth.

Or Parent-worlds and their colonies here—

Super-Romaniumus—

Or where the first Romans came from.

It's as good as the Romulus and Remus story.

Super-Israelimus—

Or that, despite modern reasoning upon this subject, there was once something that was super-parental or tutelary to early orientals.

Azuria, whence came the blue Britons, whose descendants gradually diluting, like blueing in a wash-tub, where a faucet's turned on, have been most emphasized of sub-tutelarians, or assimilators ever since.

Worlds that were once tutelarian worlds—before this earth became sole property of one of them—their attempts to convert or assimilate—but then the state that comes to all things in their missionary-frustrations—unacceptance by all stomachs of some things; rejection by all societies of some units; glaciers that sort over and cast out stones—

Repulsion. Wrath of the baffled missionary. There is no other wrath. All repulsion is reaction to the unassimilable.

So then the wrath of Azuria—

Because surrounding peoples of this earth would not assimilate with her own colonists in the part of the earth that we now call England.

I don't know that there has ever been more nearly just, reasonable, or logical wrath, in this earth's history—if there is no other wrath.

The wrath of Azuria, because the other peoples of this earth would not turn blue to suit her.

History is a department of human delusion that interest us. We are able to give a little advancement to history. In the vitrified forts of a few parts of Europe, we find data that the Humes and Gibbons have disregarded.

The vitrified forts surrounding England, but not in England.

The vitrified forts of Scotland, Ireland, Brittany, and Bohemia.

Or that, once upon a time, with electric blasts, Azuria tried to swipe this earth clear of the peoples who resisted her.

The vast blue bulk of Azuria appeared in the sky. Clouds turned green. The sun for formless and purple in the vibrations of wrath that were emanating from Azuria. The whitish, or yellowish, or brownish peoples of Scotland, Ireland, Brittany, and Bohemia fled to hilltops and built forts. In a real existence, hilltops, or easiest accessibility to an aerial enemy, would be the last choice in refuges. But here, in quasi-existence, if we're accustomed to run to hilltops, in times of danger, we run to them just the same, even with danger closest to hilltops. Very common in quasi-existence: attempt to escape by running closer to the pursuing.

They built forts, or already had forts, on hilltops.

Something poured electricity upon them.

The stones of these forts exist to this day, vitrified, or melted and turned to glass.

The archaeologists have jumped from one conclusion to another, like the "rapid chamois" we read of a while ago, to account for vitrified forts, always restricted by the commandment that unless their conclusions conformed to such tenets as Exclusionism, of the System, they would be excommunicated. So archaeologists, in their medieval dread of excommunication, have tried to explain vitrified forts in terms of terrestrial experience. We find in their insufficiencies the same old assimilating of all that could be assimilated, and disregard for the unassimilable, conventionalizing into the explanation that vitrified forts were made by prehistoric peoples who built vast fires—often remote from wood-supply—to melt externally, and to cement together, the stones of their constructions. But negativeness always: so within itself a science can never be homogeneous or unified or harmonious. So Miss Russel, in the *Journal of the B. A. A.*, has pointed out that it is seldom that single stones, to say nothing of long walls, of large houses that are burned to the ground, are vitrified.

If we pay a little attention to this subject, ourselves, before starting to write upon it, which is one of the ways of being more nearly real than oppositions so far encountered by us, we find:

That the stones of these forts are vitrified in no reference to cementing them: that they are cemented here and there, in streaks, as if special blasts had struck, or played, upon them.

Then one thinks of lightning?

Once upon a time something melted, in streaks, the stones of forts on the tops of hills in Scotland, Ireland, Brittany, and Bohemia.

Lightning selects the isolated and conspicuous.

But some of the vitrified forts are not upon tops of hills: some are very inconspicuous: their walls too are vitrified in streaks.

Something once had effect, similar to lightning, upon forts, mostly on hills, in Scotland, Ireland, Brittany, and Bohemia.

But upon hills, all over the rest of the world are remains of forts that are not vitrified.

There is only one crime, in the local sense, and that is not to turn blue, if the gods are blue: but, in the universal sense, the one crime is not to turn the gods themselves green, if you're green.

Chapter 13

One of the most extraordinary of phenomena, or alleged phenomena, of psychic research, or alleged research—if in quasi-existence there never has been real research, but only approximations to research that merge away, or that are continuous with, prejudice and convenience—

"Stone-throwing."

It's attributed to poltergeists. They're mischievous spirits.

Poltergeists do not assimilate with our own present quasi-system, which is an attempt to correlate denied or disregarded data as phenomena of extra-telluric forces, expressed in physical terms. Therefore I regard poltergeists as evil or false or discordant or absurd—names that we give to various degrees or aspects of the unassimilable, or that which resists attempts to organize, harmonize, systematize, or, in short, to positivize—names that we give to our recognitions of the negative sale. I don't care to deny poltergeists, because I suspect that later, when we're more enlightened, or when we widen the range of our credulities, or take on more of that increase of ignorance that is called knowledge, poltergeists may become assimilable. Then they'll be as reasonable as trees. By reasonableness I mean that which assimilates with a dominant force, or system, or major body of thought—which is, itself, of course, hypnosis and delusion—developing, however, in our acceptance, to higher and higher approximations to realness. The poltergeists are now evil or absurd to me, proportionately to their present unassimilableness, compounded, however, with the factor of their possible future assimilableness.

We lug in the poltergeists, because some of our own data, or alleged data, merge away indistinguishably with data, or alleged data, of them:

Instances of stones that have been thrown, or that have fallen, upon a small area, from an unseen and undetectable source.

London *Times*, April 27, 1872:

"From 4 o'clock, Thursday afternoon, until half past

eleven, Thursday night, the houses, 56 and 58 Reverdy Road, Bermondsey, were assailed with stones and other missiles coming from an unseen quarter. Two children were injured, every window broken, and several articles of furniture were destroyed. Although there was a strong body of policemen scattered in the neighbourhood, they could not trace the direction whence the stones were thrown."

"Other missiles" make a complication here. But if the expression means tin cans and old shoes, and if we accept that the direction could be traced because it never occurred to anyone to look upward—why, we've lost a good deal of our provincialism by this time.

London *Times*, Sept. 16, 1841:

That, in the home of Mrs. Charlton, at Sutton Courthouse, Sutton Lane, Chiswick, windows had been broken "by some unseen agent". Every attempt to detect the perpetrator failed. The mansion was detached and surrounded by high walls. No other building was near it.

The police were called. Two constables, assisted by members of the household, guarded the house, but the windows continued to be broken "both in front and behind the house".

Or the floating islands that are often stationary in the Super-Sargasso Sea; and atmospheric disturbances that sometimes affect them, and bring things down within small areas, upon this earth, from temporarily stationary sources.

Super-Sargasso Sea and the beaches of its floating islands from which I think, or at least, accept, pebbles have fallen:

Wolverhampton, England, June, 1860—violent storm—fall of so many little black pebbles that they were cleared away by shovelling (*La Sci. Pour Tous*, 5–264); great number of small black stones that fell at Birmingham, England, August, 1858—violent storm—said to be similar to some basalt a few leagues from Birmingham (*Rept. Brit. Assoc.*, 1864–37); pebbles described as "common water-worn pebbles" that fell at Palestine, Texas, July 6, 1888—"of a formation not found near Palestine" (W. H. Perry, Sergeant, Signal Corps, *Monthly Weather Review*, July, 1888); "a number of stones of peculiar formation and shapes, unknown in this neighbourhood, fell in a tornado at Hillsboro, Ill., May 18, 1883". (*Monthly Weather Review*, May, 1883.)

Pebbles from aerial beaches and terrestrial pebbles as

products of whirlwinds, so merge in these instances that, though it's interesting to hear of things of peculiar shape that have fallen from the sky, it seems best to pay little attention here, and to find phenomena of the Super-Sargasso Sea remote from the merger:

To this requirement we have three adaptations:

Pebbles that fell where no whirlwind to which to attribute them could be learned of;

Pebbles which fell in hail so large that incredibly could that hail have been formed in the earth's atmosphere;

Pebbles which fell and were, long afterward, followed by more pebbles, as if from some aerial, stationary source, in the same place. In September, 1898, there was a story in a New York newspaper, of lightning—or an appearance of luminosity?—in Jamaica—something had struck a tree: near the tree were found some small pebbles. It was said that the pebbles had fallen from the sky, with the lightning. But the insult to orthodoxy was that they were not angular fragments such as might have been broken from a stony meteorite: that they were "water-worn pebbles".

In the geographical vagueness of a mainland, the explanation "up from one place and down in another" is always good, and is never overworked, until the instances are massed as they are in this book: but, upon this occasion, in the relatively small area of Jamaica, there was no whirlwind findable—however "there in the first place" bobs up.

Monthly Weather Review, August, 1898–363:

That the government meteorologist had investigated: had reported that a tree had been struck by lightning, and that small water-worn pebbles had been found near the tree: but that similar pebbles could be found all over Jamaica.

Monthly Weather Review, September, 1915–446:

Prof. Fassig gives an account of a fall of hail that occurred in Maryland, June 22, 1915: hailstones the size of baseballs "not at all uncommon".

"An interesting, but unconfirmed, account stated that small pebbles were found at the centre of some of the larger hail gathered at Annapolis. The young man who related the story offered to produce the pebbles, but has not done so."

A footnote:

"Since writing this, the author states that he has received

some of the pebbles."

When a young man "produces" pebbles, that's as convincing as anything else I've ever heard of, though no more convincing than, if having told of ham sandwiches falling from the sky, he should "produce" ham sandwiches. If this "reluctance" be admitted by us, we correlate it with a datum reported by a Weather Bureau observer, signifying that, whether the pebbles had been somewhere aloft a long time or not, some of the hailstones that fell with them, had been. The datum is that some of these hailstones were composed of from 20 to 25 layers alternately of clear ice and snow-ice. In orthodox terms I argue that a fair-sized hailstone falls from the clouds with velocity to warm it so that it would not take on even one layer of ice. To put on 20 layers of ice, I conceive of something that had not fallen at all, but had rolled somewhere, at a leisurely rate, for a long time.

We now have a commonplace datum that is familiar in two respects:

Little, symmetric objects of metal that fell at Orenburg, Russia, September, 1924 (*Phil. Mag.*, 4–8–463).

I now think of the disc of Tarbes, but when first I came upon these data I was impressed only with recurrence, because the objects of Orenburg were described as crystals of pyrites, or sulphate of iron. I had no notion of metallic objects that might have been shaped or moulded by means other than crystallization, until I came to Arago's account of these occurrences (*Euvres*, 11–644). Here the analysis gives 70 per cent red oxide of iron, and sulphur and loss by ignition 5 per cent. It seems to me acceptable that iron with considerably less than 5 per cent sulphur in it is not iron pyrites—then little, rusty iron objects, shaped by some other means, have fallen, four months apart, at the same place. M. Arago expresses astonishment at this phenomenon of recurrence so familiar to us.

Altogether, I find opening before us, vistas of heresies to which I, for one, must shut my eyes. I have always been in sympathy with the dogmatists and exclusionists: that is plain in our opening lines: that to seem to be is falsely and arbitrarily and dogmatically to exclude. It is only that exclusionists who are in the nineteenth century are evil in the twentieth century. Constantly we feel a merging away into infinitude; but that this book shall approximate to form, or that our data shall approximate to organization, or that we shall approximate to intelligibility, we have to call ourselves back constantly from wandering off into infinitude. The thing that we do, however, is to make our own outline, or the difference between what we include and what we exclude, vague.

The crux here, and the limit beyond which we may not go—very much—is:

Acceptance that there is a region that we call the Super-Sargasso Sea—not yet fully accepted, but a provisional position that has received a great deal of support—

But is it a part of this earth, and does it revolve with and over this earth—

Or does it flatly overlie this earth, nor revolving with and over this earth—

That this earth does not revolve, and is not round, or roundish, at all, but is continuous with the rest of its system, so that, if one could break away from the traditions of the geographers, one might walk and walk, and come to Mars, and then find Mars continuous with Jupiter?

I suppose some day such queries will sound absurd—the thing will be so obvious—

Because it is very difficult for me to conceive of little metallic objects hanging precisely over a small town in Russia, for four months, if revolving, unattached, with a revolving earth—

It may be that something aimed at that town, and then later took another shot.

There are speculations that seem to me to be evil relatively to these years in the twentieth century—

Just now, I accept that this earth is—not round, or course: that is very old-fashioned—but roundish, or at least, that it has what is called form of its own, and does revolve upon its axis, and in an orbit around the sun. I only accept these traditional notions—

And that above it are regions of suspension that revolve with it: from which objects fall, by disturbances of various kinds, and then, later, fall again, in the same place:

Monthly Weather Review, May, 1884–134:

Report from the Signal Service observer, at Bismarck, Dakota:

That at 9 o'clock, in the evening of May 22, 1884, sharp sounds were heard throughout the city, caused by a fall of flinty stones striking against windows.

Fifteen hours later another fall of flinty stones occurred at Bismarck.

There is no report of stones having fallen anywhere else.

This is a thing of the ultra-damned. All Editors of scientific publications read the *Monthly Weather Review* and frequently copy from it. The noise made by the stones of Bismarck, rattling against those windows, may be in a language that aviators will some day interpret: but it was a noise entirely surrounded by silences. Of this ultra-damned thing, there is no mention, findable by me, in any other publication.

The size of some hailstones has worried many meteorologists—but not text-book meteorologists. I know of no more serene occupation than that of writing text-books—though writing for the *War Cry*, of the Salvation Army, may be equally unadventurous. In the drowsy tranquillity of a textbook, we easily and unintelligently read of dust particles around which icy rain forms, hailstones, in their fall then increasing by accretion—but in the meteorological journals, we read often of air-spaces nucleating hailstones—

But it's the size of the things. Dip a marble in icy water. Dip and dip it. If you're a resolute dipper, you will, after a while, have an object the size of a baseball—but I think a thing could fall from the moon in that length of time. Also the strata of them. The Maryland hailstones are unusual, but a dozen strata have often been counted. Ferrell gives an instance of thirteen strata. Such considerations led Prof. Schwedoff to argue that some hailstones are not, and cannot, be generated in this earth's atmosphere—that they come from somewhere else. Now, in a relative existence, nothing can of itself be either attractive or repulsive: its effects are functions of its associations or implications. Many of our data have been taken from very conservative scientific sources: it was not until their discordant implications, or irreconcilabilities with the System, were perceived, that excommunication was pronounced against them.

Prof. Schwedoff's paper was read before the British Association (*Rept. of 1882*, p. 453).

The implication, and the repulsiveness of the implication to the snug and tight exclusionists of 1882—though we hold out that they were functioning well and ably relatively to 1882—

That there is water—oceans or lakes and ponds, or rivers of it—that there is water away from and yet not far-remote from, this earth's atmosphere and gravitation—

The pain of it:

That the snug little system of 1882 would be ousted from its reposefulness—

A whole new science to learn:

The Science of Super-Geography—

And Science is a turtle that says that its own shell encloses all things.

So the members of the British Association. To some of them Prof. Schwedoff's ideas were like slaps on the back of an environment-denying turtle: to some of them his heresy was like an offering of meat, raw and dripping, to milk-fed lambs, and some of them turled like turtles. We used to crucify, but now we ridicule: or, in the loss of vigour of all progress, the spike has etherealized into the laugh.

Sir William Thomson ridiculed the heresy, with the phantomosities of his era:

That all bodies, such as hailstones, if away from this earth's atmosphere, would have to move at planetary velocity—which would be positively reasonable if the pronouncements of St. Isaac were anything but articles of faith—that a hailstone falling through this earth's atmosphere, with planetary velocity, would perform 13,000 times as much work as would raise an equal weight of water one degree centigrade, and therefore never fall as a hailstone at all; be more than melted—super-volatilized—

These turls and these bleats of pedantry—though we insist that, relatively to 1882, these turls and bleats should be regarded as respectfully as we regard rag dolls that keep infants occupied and noiseless—it is the survival of rag dolls into maturity that we object to—so these pious and naïve ones who believed that 13,000 times something could have—that is, in quasi-existence—an exact and calculable resultant, whereas there is—in quasi-existence—nothing that can, except by delusion and convenience, be called a unit, in the first place—whose devotions to St. Isaac required blind belief in formulas of falling bodies—

Against data that were piling up, in their own time, of slow-falling meteorites; "milk warm" ones admitted even by Farrington and Merrill; at least one icy meteorite nowhere denied by the present orthodoxy, a datum as accessible to Thomson, in 1882, as it is now to us, because it was an occurrence of 1860. Beans and needles and tacks and a magnet. Needles and tacks adhere to and systematize relatively to a magnet, but, if some beans, too, be caught up, they are irreconcilables to this system and drop right out of it. A member of the Salvation Army may hear over and over data that seem so memorable to an evolutionist. It seems remarkable that they do not influence him—one finds that he cannot remember them. It is incredible that Sir William Thomson had never heard of slow-falling, cold meteorites. It is simply that he had no power to remember such irreconcilabilities.

And then Mr. Symons again. Mr. Symons was a man who probably did more for the science of meteorology than did any other man of this time: therefore he probably did more to hold back the science of meteorology than did any other man of his time. In *Nature*, 41--135, Mr. Symons says that Prof. Schwedoff's ideas are "very droll".

I think that even more amusing is our own acceptance that, not very far above this earth's surface, is a region that will be the subject of a whole new science—super-geography—with which we shall immortalize ourselves in the resentments of the schoolboys of the future—

Pebbles and fragments of meteors and things from Mars and Jupiter and Azuria: wedges, delayed messages, cannon balls, bricks, nails, coal and coke and charcoal and offensive old cargoes—things that coat in ice in some regions and things that get into areas so warm that they putrefy—or that there are all the climates of geography in super-geography. I shall have to accept that, floating in the sky of this earth, there often are fields of ice as extensive as those on the Arctic Ocean—volumes of water in which are many fishes and frogs—tracts of land covered with caterpillars—

Aviators of the future. They fly up and up. Then they get out and walk. The fishing's good: the bait's right there. They find messages from other worlds—and within three weeks there's a big trade worked up in forged messages. Sometime I shall write a guide book to the Super-Sargasso Sea, for aviators, but just at present there wouldn't be much call for it.

We now have more of our expression upon hail as a

concomitant, or more data of things that have fallen from the sky, without hail.

In general, the expression is:

These things may have been raised from some other part of the earth's surface, in whirlwinds, or may not have fallen, and may have been upon the ground, in the first place—but were the hailstones found with them, raised from some other part of the earth's surface, or were the hailstones upon the ground, in the first place?

As I said before, this expression is meaningless as to a few instances; it is reasonable to think of some coincidence between the fall of hail and the fall of other things: but, inasmuch as there have been a good many instances,—we begin to suspect that this is not so much a book we're writing as a sanitarium for overworked coincidences. If not conceivably could very large hailstones and lumps of ice form in this earth's atmosphere, and so then had to come from external regions, then other things in or accompanying very large hailstones and lumps of ice came from external regions—which worries us a little: we may be instantly translated to the Positive Absolute.

Cosmos, 13–120, quotes a Virginia newspaper, that fishes said to have been catfishes, a foot long, some of them, had fallen, in 1853, at Norfolk, Virginia, with hail.

Vegetable débris, not only nuclear, but frozen upon the surfaces of large hailstones, at Toulouse, France, July 28, 1874. (*La Science Pour Tous*, 1874–270.)

Description of a storm, at Pontiac, Canada, July 11, 1864, in which it is said that it was not hailstones that fell, but "pieces of ice, from half an inch to over two inches in diameter" (*Canadian Naturalist*, 2–1–308):

"But the most extraordinary things is that a respectable farmer, of undoubted veracity, says he picked up a piece of hail, or ice, in the centre of which was a small green frog."

Storm at Dubuque, Iowa, June 16, 1882, in which fell hailstones and pieces of ice (*Monthly Weather Review*, June, 1882):

"The foreman of the Novelty Iron Works, of this city, states that in two large hailstones melted by him were found small living frogs." But the pieces of ice that fell upon this occasion had a peculiarity that indicates—though by as bizarre an indication as any we've had yet—that they had been for a long time motionless or floating somewhere.

We'll take that up soon.

Living Age, 52–186:

That, June 30, 1841, fishes, one of which was 10 inches long, fell at Boston; that, eight days later, fishes and ice fell at Derby.

In Timb's *Year Book* 1842–275, it is said that, at Derby, the fishes had fallen in enormous numbers; from half an inch to two inches long, and some considerably larger. In the *Athenaeum*, 1841–542, copied from the Sheffield *Patriot*, it is said that one of the fishes weighed three ounces. In several accounts, it is said that, with the fishes, fell many small frogs and "pieces of half-melted ice". We are told that the frogs and the fishes had been raised from some other part of the earth's surface, in a whirlwind; no whirlwind specified; nothing said as to what part of the earth's surface comes ice, in the month of July—interests us that the ice is described as "half-melted". In the London *Times,* July 15, 1841, it is said that the fishes were sticklebacks; that they had fallen with ice and small frogs, many of which had survived the fall. We note that, at Dunfermline, three months later (Oct. 7, 1841), fell many fishes, several inches in length, in a thunderstorm. (London *Times*, Oct. 12, 1841.)

Hailstones, we don't care so much about. The matter of stratification seems significant, but we think more of the fall of lumps of ice from the sky, as possible data of the Super-Sargasso Sea:

Lumps of ice, a foot in circumference, Derbyshire, England, May 12, 1811 (*Annual Register*, 1811–54); cuboidal mass, six inches in diameter, that fell at Birmingham, 26 days later (Thomson, *Intro. To Meteorology*, p. 179); size of pumpkins, Bungalore, India, May 22, 1851 (*Rept. Brit. Assoc.*, 1855–35); masses of ice of a pound and a half each, New Hampshire, Aug. 13, 1851 (Lummis, *Meteorology*, p. 129); masses of ice, size of a man's head, in the Delphos tornado (Ferrel, *Popular Treatise*, p. 428); Large as a man's hand, killing thousands of sheep, Texas, May 3, 1877 (*Monthly Weather Review,* May, 1877); "pieces of ice so large that they could not be grasped in one hand", in a tornado, in Colorado, June 24, 1877 (*Monthly Weather Review*, June, 1877); lumps of ice four and a half inches long, Richmond, England, Aug. 2, 1879 (*Symons' Met. Mag.*, 14–100); mass of ice, 21 inches in circumference that fell with hail, Iowa, June, 1881 (*Monthly Weather Review*, June, 1881); "Pieces of ice" eight inches long, and an inch and a half thick, Davenport, Iowa, Aug. 30, 1882 (*Monthly Weather Review*, Aug., 1882): lump of ice size of a brick; weight two pounds, Chicago, July 12, 1883 (*Monthly Weather Review*, July, 1883); lumps of ice that weighed one pound and a half each, India, May (?), 1888 (*Nature*, 37–42); lump of ice weighing four pounds, Texas, Dec. 6, 1893 (*Sc. Am.*, 68–58); lumps of ice one pound in weight, Nov. 14, 1901, in a tornado, Victoria (*Meteorology of Australia*, p. 34).

Of course it is our acceptance that these masses not only accompanied tornadoes, but were brought down to this earth by tornadoes.

Flammarion, *The Atmosphere*, p. 34:

Block of ice, weighing four and a half pounds that fell at Cazorta, Spain, June 15, 1829; block of ice, weighing 11 pounds, at Cette, France, October, 1844; mass of ice three feet long, three feet wide, and more than two feet thick, that fell, in a storm, in Hungary, May 8, 1802.

Scientific American, 47–119:

That, according to the *Salina Journal* a mass of ice weighing about 80 pounds had fallen from the sky, near Salina, Kansas, August, 1882. We are told that Mr. W. J. Hagler, the North Santa Fé merchant became possessor of it, and packed it in sawdust in his store.

London *Times*, April 7, 1860:

That, upon the 16th of March, 1860, in a snowstorm, in Upper Wasdale, blocks of ice, so large that at a distance they looked like a flock of sheep had fallen.

Rept. Brit. Assoc., 1851–32:

That a mass of ice about a cubic yard in size had fallen at Candeish, India, 1828.

Against these data, though, so far as I know, so many of them have never been assembled together before, there is a silence upon the part of scientific men that is unusual. Our Super-Sargasso Sea may not be an unavoidable conclusion, but arrival upon this earth of ice from external regions does seem to be—except that there must be, be it ever so faint, a merger. It is in the notion that these masses of ice are only congealed hailstones. We have data against this notion, as applied to all our instances but the explanation has been offered, and, it seems to me, may apply in some instances. In the *Bull. Soc. Astro. de France*, 20–245, it is said of blocks of ice the size of decanters that had fallen at Tunis that they were only masses of congealed hailstones.

London *Times*, Aug. 4, 1857:

That a block of ice, described as "pure" ice, weighing 25 pounds, had been found in the meadow of Mr. Warner, of Cricklewood. There had been a storm the day before. As in some of our other instances, no one had seen this object fall from the sky. It was found after the storm: that's all that can be said about it.

Letter from Capt. Blakiston, communicated by Gen. Royal Society (London *Roy. Soc. Proc.*, 10–468):

That, Jan. 14, 1860, in a thunderstorm, pieces of ice had fallen upon Capt. Blakiston's vessel—that it was not hail. "It was not hail, but irregular-shaped pieces of solid ice of different dimensions, up to the size of half a brick."

According to the *Advertiser-Scotsman*, quoted by the *Edinburgh New Philosophical Magazine*, 47–371, an irregular-shaped mass of ice fell at Ord, Scotland, August, 1849, after "an extraordinary peal of thunder".

It is said that this was homogeneous ice, except in a small part, which looked like congealed hailstones.

The mass was about 20 feet in circumference.

The story as told in the London *Times*, Aug. 14, 1849, is that, upon the evening of the 13th of August, 1849, after a loud peal of thunder, a mass of ice said to have been 20 feet in circumference, had fallen upon the estate of Mr. Moffat, of Balvullich, Ross-shire. It is said that this object fell alone, or without hailstones.

Altogether, though it is not so strong for the Super-Sargasso Sea, I think this is one of our best expressions upon external origins. That large blocks of ice could form in the moisture of this earth's atmosphere is about as likely as that blocks of stone could form in a dust whirl. Of course, if ice or water comes to this earth from external sources, we think of at least minute organisms in it, and on, with our data, to frogs, fishes; on to anything that's thinkable, coming from external sources. It's of great importance to us to accept that large lumps of ice have fallen from the sky, but what we desire most—perhaps because of our interest in its archaeologic and paleontologic treasures—is now to be through with tentativeness and probation, and to take the Super-Sargasso Sea into full acceptance in our more advanced fold of the chosen of this twentieth century.

In the *Report of the British Association*, 1855–37, it is said that, at Poorhundur, India, Dec. 11, 1854, flat pieces of ice, many of them weighing several pounds—each, I suppose—had fallen from the sky. They are described as "large ice-flakes".

Vast fields of ice in the Super-Arctic regions, or strata, of the Super-Sargasso Sea. When they break up, their fragments are flake-like. In our acceptance, there are aerial ice-fields that are remote from this earth; that break up, fragments grinding against one another, rolling in vapour and water, of different constituency in different regions, forming slowly as stratified hailstones—but that there are ice-fields near this earth, that break up into just such flat pieces of ice as cover any pond or river when ice of a pond or river is broken, and sometimes soon precipitated to the earth, in this familiar flat formation.

Symons' Met. Mag., 43–154:

A correspondent writes that, at Braemar, July 2, 1908, when the sky was clear overhead, and the sun shining, flat pieces of ice fell—from somewhere: thunder was heard.

Until I saw the reproduction of a photograph in the *Scientific American*, Feb. 21, 1914, I had supposed that these ice fields must be say, at least ten or twenty miles away from this earth, and invisible, to terrestrial observers, except as the blurs that have so often been reported by astronomers and meteorologists. The photograph published by the *Scientific American* is of an aggregation supposed to be clouds, presumably not very high, so clearly detailed are they. The writer says that they looked to him like "a field of broken ice". Beneath is a picture of a conventional field of ice, floating ordinarily in water. The resemblance between the two pictures is striking—nevertheless, it seems to me incredible that the first of the photographs could be of an aerial ice-field, or that gravitation could cease to act at only a mile or so from this earth's surface—

Unless:

The exceptional: the flux and vagary of all things.

Or that normally this earth's gravitation extends, say, ten or fifteen miles outward—but that gravitation must be rhythmic.

Of course, in the pseudo-formulas of astronomers, gravitation as a fixed quantity is essential. Accept that gravitation is a variable force, and astronomers deflate, with a perceptible hissing sound, into the punctured condition of economists, biologists, meteorologists, and all the others of the humbler divinities, who can admittedly offer only insecure approximations.

We refer all who would not like to hear the hiss of escaping arrogance, to Herbert Spencer's chapters upon the rhythm of all phenomena.

If everything else—light from the stars, heat from the sun, the winds and the tides; forms and colours and sizes of animals; demands and supplies and prices; political opinions and chemic reactions and religious doctrines and magnetic intensities and the ticking of clocks; and arrival and departure of the seasons—if everything else is variable, we accept that the notion of gravitation as fixed and formulable is only another attempted positivism, doomed, like all the other illusions of realness in quasi-existence. So it is intermediatism to accept that, though gravitation may approximate higher to invariability than do the winds, for instance, it must be somewhere between the Absolutes of Stability and Instability. Here then we are not much impressed with the opposition of physicists and astronomers, fearing, a little mournfully, that their language is of expiring sibilations.

So then the fields of ice in the sky, and that, though usually so far away as to be mere blurs, at times they come close enough to be seen in detail. For description of what I call a "blur", see *Pop. Sci. News*, February, 1884—sky, in general, unusually clear, but, near the sun, "a white, slightly curdled haze, which was dazzlingly bright".

We accept that sometimes fields of ice pass between the sun and the earth: that many strata of ice or very thick fields of ice, or superimposed fields would obscure the sun—that there have been occasions when the sun was eclipsed by fields of ice:

Flammarion, *The Atmosphere*, p. 394:

That a profound darkness came upon the city of Brussels, June 18, 1839:

There fell flat pieces of ice, an inch long.

Intense darkness at Aitkin, Minn., April 2, 1889: sand

and "solid chunks of ice" reported to have fallen (*Science*, April 19, 1889).

In *Symons' Meteorological Magazine*, 32–172 are outlined rough-edged but smooth-surfaced pieces of ice that fell at Manassas, Virginia, Aug. 10, 1897. They look as much like the roughly broken fragments of a smooth sheet of ice—as ever have roughly broken fragments of a smooth sheet of ice looked. About two inches across, and one inch thick. In *Cosmos*, 3–116, it is said that, at Rouen, July 5, 1853, fell irregular-shaped pieces of ice, about the size of a hand, described as looking as if all had been broken from one enormous block of ice. That I think, was an aerial iceberg. In the awful density, or almost absolute stupidity of the nineteenth century, it never occurred to anybody to look for traces of polar bears or of seals upon these fragments.

Of course, seeing what we want to see, having been able to gather these data only because they are in agreement with notions formed in advance, we are not so respectful to our own notions as to a similar impression forced upon an observer who had no theory or acceptance to support. In general, our prejudices see and our prejudices investigate, but this should not be taken as an absolute.

Monthly Weather Review, July, 1894:

That, from the Weather Bureau, of Portland, Oregon, a tornado, of June 3, 1894, was reported.

Fragments of ice fell from the sky.

They averaged three to four inches square, and about an inch thick. In length and breadth they had the smooth surfaces required by our acceptance: and, according to the writer in the *Review*, "gave the impression of a vast field of ice suspended in the atmosphere, and suddenly broken into fragments about the size of the palm of the hand".

This datum, profoundly of what we used to call the "damned," or before we could no longer accept judgment, or cut and dried condemnation by infants, turtles, and lambs, was copied—but without comment—in the *Scientific American,* 71–371.

Our theology is something like this:

Of course we ought to be damned—but we revolt against adjudication by infants, turtles, and lambs.

We now come to some remarkable data in a rather difficult department of super-geography. Vast fields of aerial ice. There's a lesson to me in the treachery of the imaginable. Most of our opposition is in the clearness with which the conventional, but impossible, becomes the imaginable, and then the resistant to modifications. After it had become the conventional with me, I conceived clearly of vast sheets of ice, a few miles above this earth—then the shining of the sun, and the ice partly melting—that note upon the ice that fell at Derby—water trickling and forming icicles upon the lower surface of the ice sheet. I seemed to look up and so clearly visualized those icicles hanging like stalactites from a flat-roofed cave, in white calcite. Or I looked up at the under side of an aerial ice-lump, and seemed to see a papillation similar to that observed by a calf at times. But then—but then—if icicles should form upon the under side of a sheet of aerial ice, that would be by falling of water toward this earth; an icicle is of course an expression of gravitation—and, if water melting from ice should fall toward this earth, why not the ice itself fall before an icicle could have time to form? Of course, in quasi-existence, where everything is a paradox, one might argue that the water falls, but the ice does not, because the ice is heavier—that is, in masses. That notion, I think, belongs in a more advanced course than we are taking at present.

Our expression upon icicles:

A vast field of aerial ice—it is inert to this earth's gravitation—but by universal flux and variation, part of it sags closer to this earth, and is susceptible to gravitation—by cohesion with the main mass, this part does not fall, but water melting from it does fall, and forms icicles—than, by various disturbances, this part sometimes falls in fragments that are protrusive with icicles.

Of the ice that fell, some of it enclosing living frogs, at Dubuque, Iowa, June 16, 1882, it is said (*Monthly Weather Review*, June, 1882) that there were pieces from one to seventeen inches in circumference, the largest weighing one pound and three-quarters—that upon some of them were icicles half an inch in length. We emphasize that these objects were not hailstones.

The only merger is that of knobby hailstones, or of large hailstones with protuberances wrought by crystallization: but that is no merger with terrestrial phenomena, and such formations are unaccountable to orthodoxy; or it is incredible that hail could so crystallize—not forming by accretion—in the fall of a few seconds. For an account of such hailstones, see *Nature*, 61–594. Note the size—"some of them the size of turkeys' eggs".

It is our expression that sometimes the icicles themselves have fallen, as if by concussion, or as if something had swept against the underside of an aerial floe, detaching its papillations.

Monthly Weather Review, June, 1889:

That, at Oswego, N.Y., June 11, 1889, according to the Turin (N.Y.) *Leader* there fell, in a thunderstorm, pieces of ice that "resembled the fragments of icicles".

Monthly Weather Review, 29–506:

That on Florence Island, St. Lawrence River, Aug. 8, 1901, with ordinary hail, fell pieces of ice "formed like icicles, the size and shape of lead pencils that had been cut into sections about three-eighths of an inch in length".

So our data of the Super-Sargasso Sea, and its Arctic region: and, for weeks at a time, an ice field may hang motionless over a part of this earth's surface—the sun has some effect upon it, but not much until late in the afternoon, I should say—part of it has sagged, but is held up by cohesion with the main mass—whereupon we have such an occurrence as would have been a little uncanny to us once upon a time—or fall of water from a cloudless sky, day after day, in one small part of this earth's surface, late in the afternoon, when the sun's rays had had time for their effects:

Monthly Weather Review, October, 1886:

That, according to the Charlotte *Chronicle*, Oct. 21, 1886, for three weeks there had been a fall of water from the sky, in Charlotte, N.C., localized in one particular spot, every afternoon, about three o'clock; that, whether the sky was cloudy or cloudless, the water or rain fell upon a small patch of land between two trees and nowhere else.

This is the newspaper account, and, as such, it seems in the depths of the unchosen, either by me or any other expression of the Salvation Army. The account by the Signal Service observer, at Charlotte, published in the *Review*, follows:

"An unusual phenomenon was witnessed on the 21st: having been informed that, for some weeks prior to date, rain had been falling daily, after 3 P.M., on a particular spot, near to trees, corner of 9th and D streets, I visited the place, and saw precipitation in the form of rain drops at 4.47 and 4.55 P.M., while the sun was shining brightly. On the 22nd, I again visited the place, and from 4.05 to 4.25 P.M., a light shower of rain fell from a cloudless sky ... Sometimes the precipitation falls over an area of half an acre, but always appears to centre on these two trees, and when lightest occurs there only."

Chapter 14

We see conventionally. It is not only that we think and act and speak and dress alike, because of our surrender to social attempt at Entity, in which we are super-cellular. We see what it is "proper" that we should see. It is orthodox enough to say that a horse is not a horse, to an infant—any more than is an orange an orange to the unsophisticated. It's interesting to walk along a street sometimes and look at things and wonder what they'd look like, if we hadn't been taught to see horses and trees and houses as horses and trees and houses I think that to super-sight they are local stresses merging indistinguishably into one another, in an all-inclusive nexus.

I think that it would be credible enough to say that many times have Monstrator and Elvera and Azuria crossed telescopic fields of vision, and were not even seen—because it wouldn't be proper to see them; it wouldn't be respectable, and it wouldn't be respectful: it would be insulting to old bones to see them; it would bring on evil influences from the relics of St. Isaac to see them.

But our data:

Of vast worlds that are orbitless, or that are navigable, or that are adrift in inter-planetary tides and currents: the data that we shall have of their approach, in modern times, within five or six miles of this earth—

But then their visits, or approaches, to other planets, or to other of the few regularized bodies that have surrendered to the attempted Entity of this solar system as a whole—

The question that we can't very well evade:

Have these other worlds, or super-constructions, ever been seen by astronomers?

I think there would not be much approximation to realness in taking refuge in the notion of astronomers who stare and squint and see only that which it is respectable and respectful to see. It is all very well to say that astronomers are hypnotics, and that an astronomer looking at the moon is hypnotized by the moon, but our acceptance is that the bodies of this present expression often visit the moon, or cross it, or are held in temporary suspension near it—then some of them must often have been within the diameter of an astronomer's hypnosis.

Our general expression:

That, upon the oceans of this earth, there are regularized vessels, but also that there are tramp vessels:

That, upon the super-ocean, there are regularized planets, but also that there are tramp worlds:

That astronomers are like mercantile purists who deny commercial vagabondage.

Our acceptance is that vast celestial vagabonds have been excluded by astronomers, primarily because their irresponsibilities are an affront to the pure and the precise, or to attempted positivism; and secondarily because thay have not been seen so very often. The planets steadily reflect the light of the sun: upon this uniformity a system that we call Primary Astronomy has been built up; but now the subject-matter of Advanced Astronomy is data of celestial phenomena that are sometimes light and sometimes dark, varying like some of the satellites of Jupiter, but with a wider range. However, light or dark, they have been seen and reported so often that the only important reason for their exclusion is—that they don't fit in.

With dark bodies that are probably external to our own solar system, I have, in the provincialism that no one can escape, not much concern. Dark bodies afloat in outer space would have been damned a few years ago, but now they're sanctioned by Prof. Barnard—and, if he says they're all right, you may think of them without the fear of doing something wrong or ridiculous—the close kinship we note so often between evil and the absurd—I suppose by the ridiculous I mean the froth of evil. The dark companion of Algol, for instance. Though that's a clear case of celestial miscegenation, the purists, or positivists, admit that's so. In the *Proceedings of the National Academy of Science*, 1915–394, Prof. Barnard writes of an object—he calls it an "object"—in Cephus. His idea is that there are dark, opaque bodies outside this solar system. But in the *Astrophysical Journal*, 1916–1, he modifies into regarding them as "dark nebulae". That's not so interesting.

We accept that Venus, for instance, has often been visited by other worlds, or by super-constructions, from which come cinders and coke and coal; that sometimes these things have reflected light and have been seen from this earth—by professional astronomers. It will be noted that throughout this chapter our data are accursed Brahmins—as, by hypnosis and inertia, we keep on and keep on saying, just as a good many of the scientists of the nineteenth century kept on and kept on admitting the power of the system that preceded them—or Continuity would be smashed. There's a big chance here for us to be instantaneously translated to the Positive Absolute—oh, well—

What I emphasize here is that our damned data are observations by astronomers of the highest standing, excommunicated by astronomers of similar standing—but backed up by the dominant spirit of their era—to which all minds had to equilibrate or be negligible, unheard, submerged. It would seem sometimes, in this book, as if our revolts were against the dogmatisms and pontifications of single scientists of eminence. This is only a convenience, because it seems necessary to personify. If we look over *Philosophical Transactions*, or the publications of the Royal Astronomical Society, for instance, we see that Herschel, for instance, was as powerless as any boy star-gazer, to enforce acceptance of any observation of his that did not harmonize with the system that was growing up as independently of him and all other astronomers, as a phase in the development of an embryo compels all cells to take on appearances concordantly with the design and the predetermined progress and schedule of the whole.

Visitors to Venus:

Evans, *Ways of the Planets*, p. 140:

That, in 1645, a body large enough to look like a satellite was seen near Venus. Four times in the first half of the eighteenth century, a similar observation was reported. The last report occurred in 1767.

A large body has been seen—seven times, according to *Science Gossip*, 1886–178—near Venus. At least one astronomer, Houzeau, accepted these observations and named that—world, planet, super-construction—"Neith". His views are mentioned "in passing, but without endorsement," in the *Trans. N.Y. Acad.*, 5–249.

Houzeau or someone writing for the magazine-section of a Sunday newspaper—outer darkness for both alike. A new satellite in this solar system might be a little disturbing—though the formulas of Laplace, which were considered final in his day, have survived the admittance of five or six hundred bodies not included in those formulas—a satellite to Venus might be a little disturbing, but would be explained—but a large body approaching a planet—staying a while—going away—coming back some other time—anchoring, as it were—

Azuria is pretty bad, but Azuria is no worse than Neith.

Astrophysical Journal, 1–127:

A light-reflecting body, or a bright spot near Mars: seen Nov. 25, 1894, by Prof. Pickering and others at the Lowell Observatory, above an illuminated part of Mars—self-luminous, it would seem—thought to have been a cloud—but estimated to have been about twenty miles away from the planet.

Luminous spot seen moving across the disc of Mercury, in 1799, by Harding and Schroeter. (*Monthly Notices of the R.A.S.*, 38–338.)

In the first Bulletin issued by the Lowell Observatory, in 1903, Prof. Lowell describes a body that was seen on the terminator of Mars, May 20, 1903. On May 27, it was "suspected". If still there, it had moved, we are told, about 300 miles—"probably a dust cloud".

Very conspicuous and brilliant spots seen on the disc of Mars, October and November, 1911. (*Popular Astronomy*, Vol. 19, No. 10.)

So one of them accepted six or seven observations that were in agreement, except that they could not be regularized, upon a world—planet—satellite—and he gave it a name. He named it "Neith".

Monstrator and Elvera and Azuria and Super-Romanimus—

Or heresy and orthodoxy and the oneness of all quasi-ness, and our ways and means and methods are the very same. Or, if we name things that may not be, we are not of lonely guilt in the nomenclature of absences—

But now Leverrier and "Vulcan".

Leverrier again.

Or to demonstrate the collapsibility of a froth, stick a pin in the largest bubble of it. Astronomy and inflation: and by inflation we mean expansion of the attenuated. Or that the science of Astronomy is a phantom-film distended with myth-stuff—but always our acceptance that it approximates higher to substantiality than did the system that preceded it.

So Leverrier and the "planet Vulcan".

And we repeat, and it will do us small good to repeat. If you be of the masses that the astronomers have hypnotized —being themselves hypnotized, or they could hypnotize others —or that the hypnotist's control is not the masterful power that it is popularly supposed to be, but only transference of state from one hypnotic to another—

If you be of the masses that the astronomers have hypnotized, you will not be able even to remember. Ten pages from here, and Leverrier and the "planet Vulcan" will have fallen from your mind, like beans from a magnet, or like data of cold meteorites from the mind of a Thomson.

Leverrier and the "planet Vulcan".

And much the good it will do us to repeat.

But at least temporarily we shall have an impression of a historic fiasco, such as, in our acceptance, could occur only in a quasi-existence.

In 1859, Dr. Lescarbault, an amateur astronomer, of Orgères, France, announced that, upon March 26, of that year, he had seen a body of planetary size cross the sun. We are in a subject that is now as unholy to the present system as ever were its own subjects to the system that preceded it, or as ever were slanders against miracles to the preceding system. Nevertheless few text-books go so far as quite to disregard this tragedy. The method of the system-artists is slightingly to give a few instances of the unholy, and dispose of the few. If it were desirable to them to deny that there are mountains upon this earth, they would record a few observations upon some slight eminences near Orange, N.J. but, say that commuters, though estimable persons in several ways, are likely to have their observations mixed. The text-books casually mention a few of the "supposed" observations upon "Vulcan", and then pass on.

Dr. Lescarbault wrote to Leverrier, who hastened to Orgères—

Because this announcement assimilated with his own calculations upon a planet Mercury and the sun—

Because this solar system itself has never attained positiveness in the aspect of Regularity: there are to Mercury, as there are to Neptune, phenomena irreconcilable with the formulas, or motions that betray influence by something else.

We are told that Leverrier "satisfied himself as to the substantial accuracy of the reported observation". The story of this investigation is told in *Monthly Notices*, 20–98. It seems too bad to threaten the naïve little thing with our rude sophistications, but it is amusingly of the ingenuousness of the age from which dogmas have survived. Lescarbault wrote to Leverrier. Leverrier hastened to Orgères. But he was careful not to tell Lescarbault who he was. Went right in and "subjected Dr. Lescarbault to a very severe cross-examination"—just as anyone might go into someone else's house and press him hard, though unknown to the hard-pressed one. Not until he was satisfied, did Leverrier reveal his identity. I suppose Dr. Lescarbault expressed astonishment. I think there's something Utopian about this: it's so unlike the stand-offishness of New York life.

Leverrier gave the name "Vulcan" to the object that Dr. Lescarbault had reported.

By the same means by which he is even to this day, supposed—by the faithful—to have discovered Neptune, he had already announced the probable existence of an Intra-Mercurial body, or group of bodies. He had five observations besides Lescarbault's upon something that had been seen to cross the sun. In accordance with the mathematical hypnoses of his era, he studied these six transits. Out of them he computed elements giving "Vulcan" a period of about 20 days, or a formula for heliocentric longitude at any time.

But he placed the time of best observation away up in 1877.

But even so, or considering that he still had probably a good many years to live, it may strike one that he was a little rash—that is if one has not gone very deep into the study of hypnoses—that, having "discovered" Neptune by a method which, in our acceptance, had no more to recommend it than had once equally well-thought-of-methods of witch-finding, he should not have taken such chances: that if he was right as to Neptune, but should be wrong as to "Vulcan," his average would be away below that of most fortune-tellers, who could scarcely hope to do business upon a 50 per cent basis—all that the reasoning of a tyro in hypnoses.

208

The date:
March 22, 1877.

The scientific world was up on its hind legs nosing the sky. The thing had been done so authoritatively. Never a pope had said a thing with more of the seeming of finality. If six observations correlated, what more could be asked? The Editor of *Nature*, a week before the predicted event, though cautious, said that it is difficult to explain how six observers, unknown to one another, could have data that could be formulated, if they were not related phenomena.

In a way, at this point occurs the crisis of our whole book.

Formulas are against us.

But can astronomic formulas, backed up by observations in agreement, taken many years apart, calculated by a Leverrier, be as meaningless, in a positive sense, as all other quasi-things that we have encountered so far?

The preparations they made, before March 22, 1877. In England, the Astronomer Royal made it the expectation of his life: notified observers at Madras, Melbourne, Sydney, and New Zealand, and arranged with observers in Chile and the United States. M. Struve had prepared for observations in Siberia and Japan—

March 22, 1877—

Not absolutely, hypocritically, I think it's pathetic, myself. If anyone should doubt the sincerity of Leverrier, in this matter, we note, whether it has meaning or not, that a few months later he died.

I think we'll take up Monstrator, though there's so much to this subject that we'll have to come back.

According to the *Annual Register*, 9–120, upon the 9th of August, 1762, M. de Rostan, of Basle, France, was taking altitudes of the sun, at Lausanne. He saw a vast spindle-shaped body, about three of the sun's digits in breadth and nine in length, advancing slowly across the disc of the sun, or "at no more than half the velocity with which the ordinary solar spots move". It did not disappear until the 7th of September, when it reached the sun's limb. Because of the spindle-like form, I incline to think of a super-Zeppelin, but another observation, which it seems to indicate that it was a world, is that, though it was opaque, and "eclipsed the sun", it had around it a kind of nebulosity—or atmosphere? A penumbra would ordinarily be a datum of a sun spot, but there are observations that indicate that this object was it a considerable distance from the sun:

It is recorded that another observer, at Paris, watching the sun, at this time, had not seen this object;

But that, M. Croste, at Sole, about forty-five German leagues northward from Lausanne, had seen it, describing the same spindle-form, but disagreeing a little as to breadth. Then comes the important point: that he and M. de Rostan did not see it upon the same part of the sun. This, then, is parallax, and, compounded with invisibility at Paris, is great parallax—or that, in the course of a month, in the summer of 1762, a large opaque, spindle-shaped body traversed the disc of the sun, but at a great distance from the sun. The writer in the *Register* says: "In a word, we know of nothing to have recourse to, in the heavens, by which to explain this phenomenon." I suppose he was not a hopeless addict to explaining. Extraordinary—we fear he must have been a man of loose habits in some other respects.

As to us—

Monstrator.

In the *Monthly Notices of the R.A.S.*, February, 1877, Leverrier, who never lost faith, up to the last day, gives the six observations upon an unknown body of planetary size, that he had formulated:

Fritsche, Oct. 10, 1802; Stark, Oct. 9, 1819; De Cuppis, Oct. 30, 1839; Sidebotham, Nov. 12, 1849; Lescarbault, March 26, 1859; Lummis, March 20, 1862.

If we weren't so accustomed to Science in its essential aspect of Disregard, we'd be mystified and impressed, like the Editor of *Nature*, with the formulation of these data: agreement of so many instances would seem incredible as a coincidence: but our acceptance is that, with just enough disregard, astronomers and fortune-tellers can formulate anything—or we'd engage, ourselves, to formulate periodicities in the crowds in Broadway—say that every Wednesday morning, a tall man, with one leg and a black eye, carrying a rubber plant, passes the Singer Building, at quarter past ten o'clock. Of course it couldn't really be done, unless such a man did have such periodicity, but if some Wednesday mornings it should be a small child lugging a barrel, or a fat Negress with a week's wash, by ordinary disregard that would be prediction good enough for the kind of quasi-existence we're in.

So whether we accuse, or whether we think that the word "accuse" over-dignifies an attitude toward a quasi-astronomer, or mere figment in a super-dream, our acceptance is that Leverrier never did formulate observations.

That he picked out observations that could be formulated—

That of this type are all formulas—

That, if Leverrier had not been himself helplessly hypnotized, or if he had had in him more than a tincture of realness, never could he have been beguiled by such a quasi-process: but that he was hypnotized, and so extended, or transferred, his condition to others, that upon March 22, 1877, he had this earth bristling with telescopes, with the rigid and almost inanimate forms of astronomers behind them—

And not a blessed thing of any unusuality was seen upon that day or succeeding days.

But that the science of Astronomy suffered the slightest in prestige?

It couldn't. The spirit of 1877 was behind it. If, in an embryo, some cells should not live up to the phenomena of their era, the others will sustain the scheduled appearances. Not until an embryo enters the mammalian stage are cells of the reptilian stage false cells.

It is our acceptance that there were many equally authentic reports upon large planetary bodies, that had been seen near the sun; that, of many, Leverrier picked out six; then deciding that all the other observations related to still other large planetary bodies, but arbitrarily, or hypnotically, disregarding—or heroically disregarding—every one of them—that to formulate at all he had to exclude falsely. The dénouement killed him, I think. I'm not at all inclined to place him with the Grays and Hitchcocks and Symonses. I'm not, because, though it was rather unsportsmanlike to put the date so far ahead, he did give a date, and he did stick to it with such a high approximation—

I think Leverrier was translated to the Positive Absolute.

The disregarded:

Observation, of July 26, 1819, by Gruthinson—but that was of two bodies that crossed the sun together—

Nature, 14-469:

That, according to the astronomer, J. R. Hind, Benjamin Scott, City Chamberlain of London, and Mr. Wray, had, in 1847, seen a body similar to "Vulcan" cross the sun.

Similar observation by Hind and Lowe, March 12, 1849 (*L'Année Scientifique*, 1876–9).

Nature, 14–505:

Body of apparent size of Mercury, seen, Jan. 29, 1860, by F. A. R. Russell and four other observers, crossing the sun.

De Vico's observation of July 12, 1837 (*Observatory*, 2–424).

L'Année Scientifique, 1865–16:

That another amateur astronomer, M. Coumbray, of Constantinople, had written to Leverrier, that, upon the 8th of March, 1865, he had seen a black point, sharply outlined, traverse the disc of the sun. It detached itself from a group of sun spots near the limb of the sun, and took 48 minutes to reach the other limb. Figuring upon the diagram sent by M. Coumbray, a central passage would have taken a little more than an hour. This observation was disregarded by Leverrier, because his formula required about four times that velocity. The point here is that these other observations are as authentic as those that Leverrier included; that, then, upon data as good as the data of "Vulcan", there must be other "Vulcans"—the heroic and defiant disregard, then, of trying to formulate one, omitting the others, which, by orthodox doctrine, must have influenced it greatly, if all were in the relatively narrow space between Mercury and the sun.

Observation upon another such body of April 4, 1876, by M. Weber, of Berlin. As to this observation, Leverrier was informed by Wolf, in August, 1876 (*L'Année Scientifique*, 1876–7). It made no difference, so far as can be known, to this notable positivist.

Two other observations noted by Hind and Denning—London *Times*, Nov. 3, 1871, and March 26, 1873.

Monthly Notices of the R.A.S., 20–100:

Standacher, February, 1762; Lichtenberg, Nov. 19, 1762; Hoffman, May, 1764; Dangos, Jan. 18, 1798; Stark, Feb. 12, 1820. An observation by Schmidt, Oct. 11, 1847, is said to be doubtful: but, upon page 192, it is said that this doubt had arisen because of a mistaken translation, and two other observations by Schmidt are given: Oct. 4, 1849, and Feb. 18, 1850—also an observation by Loff, Jan. 6, 1818. Observation by Steinheibel, at Vienna, April 27, 1820

(*Monthly Notices*, 1862).

Haase had collected reports of twenty observations like Lescarbault's. The list was published in 1872, by Wolf. Also there are other instances like Gruthinsen's:

Amer. Jour. Sci., 2–28–446:

Report by Pastorff that he had seen twice in 1836, and once in 1837, two round spots of unequal size moving across the sun, changing position relatively to each other, and taking a different course, if not orbit, each time: that, in 1834, he had seen similar bodies pass six times across the sun, looking very much like Mercury in his transits.

March 22, 1876—

But to point out Leverrier's poverty-stricken average—or discovering planets upon a 50 per cent. basis—would be to point out the low percentage of realness in the quasi-myth-stuff of which the whole system is composed. We do not accuse the text-books of omitting this fiasco, but we do note that there is the conventional adaptation here of all beguilers who are in difficulties—

The diverting of attention.

It wouldn't be possible in a real existence, with real mentality, to deal with, but I suppose it's good enough for the quasi-intellects that stupefy themselves with text-books. The trick here is to gloss over Leverrier's mistake, and blame Lescarbault—he was only an amateur—had delusions. The reader's attention is led against Lescarbault by a report from M. Lisas, director of the Brazilian Coast Survey, who, at the time of Lescarbault's "supposed" observation had been watching the sun in Brazil, and, instead of seeing even ordinary sun spots, had noted that the region of the "supposed transit" was of "uniform intensity".

But the meaninglessness of all utterances in quasi-existence—

"Uniform intensity" turns our way as much as against us—or some day some brain will conceive a way of beating Newton's third law—if every reaction, or resistance, is, or can be, interpretable as stimulus instead of resistance—if this could be done in mechanics, there's a way open here for someone to own the world—specifically in this matter, "uniform intensity" means that Lescarbault saw no ordinary sun spot, just as much as it means that no spot at all was seen upon the sun. Continuing the interpretation of a resistance as an assistance, which can always be done with mental forces—making us wonder what applications could be made with steam and electric forces—we point out that invisibility in Brazil means parallax quite as truly as it means absence, and inasmuch as "Vulcan" was supposed to be distant from the sun, we interpret denial as corroboration—method of course of every scientist, politician, theologian, high-school debater.

So the text-books, with no especial cleverness, because no especial cleverness is needed, lead the reader into contempt for the amateur of Orgères, and forgetfulness of Leverrier—and some other subject is taken up.

But our own acceptance:

That these data are as good as ever they were;

That, if someone of eminence should predict an earthquake, and if there should be no earthquake at the predicted time, that would discredit the prophet, but data of past earthquakes would remain as good as ever they had been. It is easy enough to smile at the illusion of a single amateur—

The mass-formation:

Fritsche, Stark, De Cuppis, Sidebotham, Lescarbault, Lummis, Gruthinson, De Vico, Scott, Wray, Russell, Hind, Lowe, Coumbray, Weber, Standacher, Lichtenberg, Dangos, Hoffman, Schmidt, Lofft, Steinheibel, Pastorff—

These are only the observations conventionally listed relatively to an Intra-Mercurial planet. They are formidable enough to prevent our being diverted, as if it were all the dream of a lonely amateur—but they're a mere advance-guard. From now on other data of large celestial bodies, some dark and some reflecting light, will pass and pass and keep on passing—

So that some of us will remember a thing or two, after the procession's over—possibly.

Taking up only one of the listed observations—

Or our impression that the discrediting of Leverrier has nothing to do with the acceptability of these data:

In the London *Times*, Jan. 10, 1860, is Benjamin Scott's account of his observation:

That, in the summer of 1847, he had seen a body that had seemed to be the size of Venus, crossing the sun. He says that, hardly believing the evidence of his sense of sight, he had looked for someone, whose hopes and ambitions would not make him so subject to illusion. He had told his little son, aged five years, to look through the telescope. The child had exclaimed that he had seen "a little balloon" crossing the sun. Scott says that he had not had sufficient self-reliance to make public announcement of his remarkable observation at the time, but that, in the evening of the same day, he had told Dr. Dick, F.R.A.S., who had cited other instances. In *The Times*, Jan. 12, 1860, is published a letter from Richard Abbott, F.R.A.S.: that he remembered Mr. Scott's letter to him upon this observation, at the time of the occurrence.

I suppose that, at the beginning of this chapter, one had the notion that, by hard scratching through musty old records we might rake up vague, more than doubtful data, distortable into what's called evidence of unrecognized worlds or constructions of planetary size—

But the high authenticity and the support and the modernity of these of the accursed that we are now considering—

And our acceptance that ours is a quasi-existence, in which above all other things, hopes, ambitions, emotions, motivations, stands Attempt to Positivize: that we are here considering an attempt to systematize that is sheer fanaticism in its disregard of the unsystematizable—that it represented the highest good in the nineteenth century—that it is mono-mania, but heroic mono-mania that was quasi-divine in the nineteenth century—

But that this isn't the nineteenth century.

As a double-sponsored Brahmin—in regard of Baptists—the objects of July 29, 1878, stand out and proclaim themselves so that nothing but disregard of the intensity of mono-mania can account for their reception by the system:

Or the total eclipse of July 29, 1878, and the reports of Prof. Watson, from Rawlins, Wyoming, and by Prof. Swift, from Denver, Colorado: that they had seen two shining objects at a considerable distance from the sun.

It's quite in accord with our general expression: not that there is an Intra-Mercurial planet, but that there are different bodies, many vast things; near this earth sometimes, near the sun sometimes; orbitless worlds, which, because of scarcely any data of collisions, we think of as under navigable control—or dirigible super-constructions.

Prof. Watson and Prof. Swift published their

observations.

Then the disregard that we cannot think of in terms of ordinary, sane exclusions.

The text-books systematists begin by telling us that the trouble with these observations is that they disagree widely: there is considerable respectfulness, especially for Prof. Swift, but we are told that by coincidence these two astronomers, hundreds of miles apart, wereilluded: their observations were so different—

Prof. Swift (*Nature*, Sept. 19, 1878):

That his own observation was "in close approximation to that given by Prof. Watson".

In the *Observatory*, 2–161, Swift says that his observations and Watson's were "confirmatory of each other".

The faithful try again:

That Watson and Swift mistook stars for other bodies.

In the *Observatory*, 2–193, Prof. Watson says that he had previously committed to memory all stars near the sun, down to the seventh magnitude—

And he's damned anyway.

How such exclusions work out is shown by Lockyer (*Nature*, Aug. 20, 1878). He says: "There is little doubt that an Intra-Mercurial planet has been discovered by Prof. Watson."

That was before excommunication was pronounced.

He says:

"If it will fit one of Leverrier's orbits"—

It didn't fit.

In *Nature*, 21–301, Prof. Swift says:

"I have never made a more valid observation, nor one more free from doubt."

He's damned anyway.

We shall have some data that will not live up to most requirements, but, if anyone would like to read how carefully and minutely these two sets of observations were made, see Prof. Swift's detailed description in the *Am. Jour. Sci.*, 116–313, and the technicalities of Prof. Watson's observations in *Monthly Notices*, 38–525.

Our own acceptance upon dirigible worlds, which is assuredly enough, more nearly real than attempted concepts of large planets relatively near this earth, moving in orbits, but visible only occasionally; which more nearly approximates to reasonableness than does wholesale slaughter of Swift and Watson and Fritsche and Stark and De Cuppis—but our own acceptance is so painful to so many minds that, in another of the charitable moments that we have now and then for the sake of contrast, we offer relief:

The things seen high in the sky by Swift and Watson—

Well, only two months before—the horse and the barn—

We go on with more observations by astronomers, recognizing that it is the very thing that has given them life, sustained them, held them together, that has crushed all but the quasi-gleam of independent life out of them. Were they not systematized, they could not be at all, except sporadically and without sustenance. They are systematized: they must not vary from the conditions of the system: they must not break away for themselves.

The two great commandments:

Thou shalt not break Continuity;

Thou shalt try.

We go on with these disregarded data, some of which, many of which, are of the highest degree of acceptability. It is the System that pulls back its variations, as this earth is pulling back the Matterhorn. It is the System that nourishes and rewards, and also freezes out life with the chill of disregard. We do note that, before excommunication is pronounced, orthodox journals do liberally enough record unassimilable observations.

All things merge away into everything else.

That is Continuity.

So the System merges away and evades us when we try to focus against it.

We have complained a great deal. At least we are not so dull as to have delusion that we know just exactly what it is that we are complaining about. We speak seemingly definitely enough of "the System", but we're building upon observations by members of that very system. Or what we are doing—gathering up the loose heresies of the orthodox. Of course "the System" fringes and ravels away, having no real outline. A Swift will antagonize "the System", and a Lockyer will call him back; but, then, a Lockyer will vary with a "meteoric hypothesis", and a Swift will, in turn, represent "the System". This state is to us typical of all intermediatist phenomena; or that not conceivably is anything really anything, if its parts are

217

likely to be their own opposites at any time. We speak of astronomers—as if there were real astronomers—but who have lost their identity in a System—as if it were a real System—but behind that System is plainly a rapport, or loss of identity in the Spirit of an Era.

Bodies that have looked like dark bodies, and lights that may have been sunlight reflected from interplanetary—objects, masses, constructions—

Lights that have been seen upon—or near?—the moon:

In *Philosophical Transactions*, 82–27, is Herschel's report upon many luminous points, which he saw, upon—or near?—the moon, during an eclipse. Why they should get luminous, whereas the moon itself was dark, would get us into a lot of trouble—except that later we shall, or we shan't, accept that many times have luminous objects been seen close to this earth—at night.

But numerousness is a new factor, or new disturbance, to our explorations—

A new aspect of inter-planetary inhabitancy or occupancy—

Worlds in hordes—or beings—winged beings perhaps—wouldn't astonish me if we should end up by discovering angles—or beings in machines—argosies of celestial voyagers—

In 1783 and 1787, Herschel reported more lights on or near the moon, which he supposed were volcanic.

The world of a Herschel has had no more weight, in divergences from the orthodox, than has had the word of a Lescarbault. These observations are of the disregarded.

Bright spots seen on the moon, November, 1821 (*Proc. London Roy. Soc.*, 2–167).

For our other instances, see Loomis (*Treatise on Astronomy*, p. 174).

A moving light is reported in *Phil. Trans.*, 84–429. To the writer, it looked like a star passing over the moon—"which, on the next moment's consideration I knew to be impossible". "It was a fixed, steady light upon the dark part of the moon." I suppose "fixed" applies to lustre.

In the *Report of the Brit. Assoc.*, 1847–18, there is an observation by Rankin, upon luminous points seen on the shaded part of the moon, during an eclipse. They seemed to this observer like reflections of stars. That's not very reasonable: however, we have, in the *Annual Register*,

1821–687, a light not referable to a star—because it moved with the moon: was seen three nights in succession; reported by Capt. Kater. See *Quart. Jour. Roy. Inst.*, 12–133.

Phil. Trans., 112–237:

Report from the Cape Town Observatory: a whitish spot on the dark part of the moon's limb. Three smaller lights were seen.

The call of positiveness, in its aspects of singleness, or homogeneity, or oneness, or completeness. In data now coming, I feel it myself. A Leverrier studies more than twenty observations. The inclination is irresistible to think that they all relate to one phenomenon. It is an expression of cosmic inclination. Most of the observations are so irreconcilable with any acceptance other than of orbitless, dirigible worlds that he shuts his eyes to more than two-thirds of them; he picks out six that can give him the illusion of completeness, or of all relating to one planet.

Or let it be that we have data of many dark bodies—still do we incline almost irresistibly to think of one of them as the dark-body-in-chief. Dark bodies, floating, or navigating, in interplanetary space—and I conceive of one that's the Prince of Dark Bodies:

Melanicus.

Vast dark thing with wings of a super-bat, or jet-black super-construction; most likely one of the spores of the Evil One.

The extraordinary year, 1883:

London, *Times*, Dec. 17, 1883:

Extract from a letter by Hicks Pashaw: that, in Egypt, Sept. 24, 1883, he had seen, through glasses, "an immense black spot on the lower part of the sun".

Sun spot, maybe.

One night an astronomer was looking up at the sky, when something obscured a star, for three and a half seconds. A meteor had been seen nearby, but its train had been only momentarily visible. Dr. Wolf was the astronomer (*Nature*, 86–528).

The next datum is one of the most sensational we have, except that there is very little to it. A dark object that was seen by Prof. Heis, for eleven degrees of arc, moving slowly across the Milky Way. (Greg's Catalogue, *Rept. Brit.*

219

One of our quasi-reasons for accepting that orbitless worlds are dirigible is the almost complete absence of data of collisions: of course, though in defiance of gravitation, they may, without direction like human direction, adjust to one another in the way of vortex rings of smoke—a very human-like way, that is. But in *Knowledge*, February, 1894, are two photographs of Brooks' comet that are shown as evidence of its seeming collision with a dark object, October, 1893. Our own wording is that it "struck against something"; Prof. Barnard's is that it had "entered some dense medium, which shattered it". For all I know it had knocked against merely a field of ice.

Melanicus.

That upon the wings of a super-bat, he broods over this earth and over other worlds, perhaps deriving something from them: hovers on wings, or wing-like appendages, or planes that are hundreds of miles from tip to tip—a super-evil thing that is exploiting us. By Evil I mean that which makes us useful.

He obscures a star. He shoves a comet. I think he's a vast, black, brooding vampire.

Science, July 31, 1896:

That, according to a newspaper account, Mr. W. R. Brooks, director of the Smith Observatory, had seen a dark round object pass rather slowly across the moon, in a horizontal direction. In Mr. Brooks' opinion it was a dark meteor. In *Science*, Sept. 14, 1896, a correspondent writes that, in his opinion, it may have been a bird. We shall have no trouble with the meteor and bird mergers, if we have observations of long duration and estimates of size up to hundreds of miles. As to the body that was seen by Brooks, there is a note from the Dutch astronomer, Muller, in the *Scientific American,* 75–251, that, upon April 4, 1892, he had seen a similar phenomenon. In *Science Gossip*, n.s., 3–135, are more details of the Brooks object—apparent diameter about one-thirtieth of the moon's—moon's disc crossed in three or four seconds. The writer, in *Science Gossip*, says that, on June 27, 1896, at one o'clock in the morning, he was looking at the moon with a 2-inch achromatic, power 44, when a long black object sailed past, from west to east, the transit occupying 3 or 4 seconds. He believed this object to be a bird—there was, however no fluttering motion observable in it.

In the *Astronomische Nachrichten*, No. 3477, Dr. Brendel, of Griefswald, Pomerania, writes that Postmaster Ziegler and other observers had seen a body about six feet in diameter crossing the sun's disc. The duration here indicates something far from the earth, and also far from the sun. This thing was seen a quarter of an hour before it reached the sun. Time in crossing the sun was about an hour. After leaving the sun it was visible an hour.

I think he's a vast, black vampire that sometimes broods over this earth and other bodies.

Communication from Dr. F. B. Harris (*Popular Astronomy*, 20–398):

That, upon the evening of Jan. 27, 1912, Dr. Harris saw, upon the moon, "an intensely black object". He estimated it to be 250 miles long and 50 miles wide. "The object resembled a crow poised, as near as anything." Clouds then cut off observation.

Dr. Harris writes:

"I cannot think that a very interesting and curious phenomenon happened."

Chapter 15

Short chapter coming now, and it's the worst of them all. I think it's speculative. It's a lapse from the usual pseudo-standards. I think it must mean that the preceding chapter was very efficiently done, and that now by the rhythm of all quasi-things—which can't be real things, if they're rhythms, because a rhythm is an appearance that turns into its own opposite and then back again—but now, to pay up, we're what we weren't. Short chapter, and I think we'll fill in with several points in Intermediatism.

A puzzle:

If it is our acceptance that, out of the Negative Absolute, the Positive Absolute is generating itself, recruiting, or maintaining, itself, via a third state, or our own quasi-state, it would seem that we're trying to conceive of Universalness manufacturing more Universalness from Nothingness. Take that up yourself, if you're willing to run the risk of disappearing with such velocity that you'll leave an incandescent train behind, and risk being infinitely happy for ever, whereas you probably don't want to be happy—I'll side-step that myself, and try to be intelligible by regarding the Positive Absolute from the aspect of Realness instead of Universalness, recalling that by both Realness and Universalness we mean the same state, or that which does not merge away into something else, because there is nothing else. So the idea is that out of Unrealness, instead of Nothingness, Realness, instead of Universalness, is, via our own quasi-state, manufacturing more Realness. Just so, but in relative terms, of course all imaginings that materialize into machines or statues, buildings, dollars, paintings or books in paper and ink are graduations from unrealness to realness—in relative terms. It would seem that that Intermediateness is a relation between the Positive Absolute and the Negative Absolute. But the absolute cannot be the related—or course a confession that we can't really think of it at all, if here we think of a limit to the unlimited. Doing the best we can, and encouraged by the reflection that we can't do worse than has been done by metaphysicians in the past, we accept that the absolute can't be the related. So then that our quasi-state is

222

not a real relation, if nothing in it is unreal. It seems think-able that the Positive Absolute can, by means of Intermedi-ateness, have a quasi-relation, or be only quasi-related, or be the unrelated, in final terms, or, at least, not be the related in final terms.

As to free will and Intermediatism—same answer as to everything else. By free will we mean Independence—or that which does not merge away into something else—so, in Intermediateness, neither free-will nor slave-will—but a different approximation for every so-called person toward one or the other of the extremes. The hackneyed way of expressing this seems to me to be acceptable way, if in Intermediateness, there is only the paradoxical: that we're free to do what we have to do.

I am not convinced that we make a fetish of the prepos-terous. I think our feeling is that in first gropings there's no knowing what will afterward be the acceptable. I think that if an early biologist heard of birds that grow on trees, he should record that he had heard of birds that grow on trees: then let sorting over of data occur afterward. The one thing that we try to tone down but that is to a great degree unavoidable is having our data all mixed up like Long Island and Florida in the minds of early American explor-ers. My own notion is that this whole book is very much like a map of North America in which the Hudson River is set down as a passage leading to Siberia. We think of Monstrator and Melanicus and of a world that is now in communication with this earth: if so, secretly, with certain esoteric ones upon this earth. Whether that world's Monstrator and Monstrator's Melanicus—must be the sub-ject of later inquiry. It would be a gross thing to do; solve up everything now and leave nothing to our disciples.

I have been very much struck with phenomena of "cup marks".

They look to me like symbols of communication.

But they do not look to me like means of communication between some of the inhabitants of this earth and other inhabitants of this earth.

My own impression is that some external force has marked, with symbols, rocks of this earth, from far away.

I do not think that cup marks are inscribed communi-cations among different inhabitants of this earth, because it seems too unacceptable that inhabitants of China, Scotland,

...nd America should all have conceived of the same system.

Cup marks are strings of cup-like impressions in rocks. Sometimes there are rings around them and sometimes they have only semi-circles. Great Britain, America, France, Algeria, Circassia, Palestine: they're virtually everywhere—except in the far north, I think. In China, cliffs are dotted with them. Upon a cliff near Lake Como, there is a maze of these markings. In Italy and Spain and India they occur in enormous numbers.

Given that a force, say, like electric force, could, from a distance, mark such a substance as rocks, as, from a distance of hundreds of miles, selenium can be marked by telephotographers—but I am of two minds—

The Lost Explorers from Somewhere, and an attempt, from Somewhere, to communicate with them: so a frenzy of showering of messages toward this earth, in the hope that some of them would mark rocks near the lost explorers—

Or that somewhere upon this earth, there is an especial rocky surface, or receptor, or polar construction, or a steep, conical hill, upon which for ages have been received messages from some other world; but that at times messages go astray and mark substances perhaps thousands of miles from the receptor;

That perhaps forces behind the history of this earth have left upon the rocks of Palestine and England and India and China records that may some day be deciphered, of their misdirected instructions to certain esoteric ones—Order of the Freemasons—the Jesuits—

I emphasize the row-formation of cup marks.

Prof. Douglas (*Saturday Review*, Nov. 24, 1883):

"Whatever may have been their motive, the cup-markers showed a decided liking for arranging their sculpturings in regularly spaced rows."

That cup marks are an archaic form of inscription was first suggested by Canon Greenwell many years ago. But more specifically adumbratory to our own expression are the observations of Rivett-Carnac (*Jour. Roy. Asiatic Soc.*, 1903–515):

That the Braille system of raised dots is an inverted arrangement of cup marks: also that there are strong resemblances to the Morse code. But no tame and systematized archaeologist can do more than casually point out resemblances, and merely suggest that strings of cup marks look like messages, because—China, Switzerland,

Algeria, America—if messages they be, there seems to be no escape from attributing one origin to them—then, if messages they be, I accept one external origin, to which the whole surface of this earth was accessible, for them.

Something else that we emphasize:

That rows of cup marks have often been likened to foot-prints.

But, in this similitude, their unilinear arrangement must be disregarded—of course often they're mixed up in every way, but arrangement in single lines is very common. It is odd that they should so often be likened to footprints: I suppose there are exceptional cases, but unless it's something that hops on one foot, or a cat going along a narrow fence-top, I don't think of anything that makes footprints one directly ahead of another—Cop, in a station house, walking a chalk line, perhaps.

Upon the Witch's Stone, near Ratho, Scotland, there are twenty-four cups, varying in size from one and a half to three inches in diameter, arranged in approximately straight lines. Locally it is explained that these are tracks of dogs' feet (*Proc. Soc. Antiq. Scotland*, 2–4–79). Similar marks are scattered bewilderingly all around the Witch's Stone—like a frenzy of telegraphing, or like messages repeating and repeating, trying to localize differently.

In Inverness-shire, cup marks are called "fairies' foot-marks". At Valna's church, Norway, and St. Peter's, Ambleteuse, there are such marks, said to be horses' hoofprints. The rocks of Clare, Ireland, are marked with prints supposed to have been made by a mythical cow (*Folklore*, 21–184).

We now have such a ghost of a thing that I'd not like to be interpreted as offering it as a datum: it simply illustrates what I mean by the notion of symbols like cups, or like footprints, which, if I like those of horses or cows, are the reverse of, or the negatives of, cups—of symbols that are regularly received somewhere upon this earth—steep, conical hill, somewhere, I think—but that have often alighted in wrong places—considerably to the mystification of persons waking up some morning to find them upon formerly blank spaces.

An ancient record—still worse, an ancient Chinese record—of a courtyard of a palace—dwellers of the palace waking up one morning, finding the courtyard marked with tracks like the footprints of an ox—supposed that the devil did it. (*Notes and Queries*, 9–6–225.)

Chapter 16

Angels.

Hordes upon hordes of them.

Beings massed like the clouds of souls, or the commingling whiffs of spirituality, or the exhalations of souls that Doré-pictured so often.

It may be that the Milky Way is a composition of stiff, frozen, finally static, absolute angels. We shall have data of little Milky Ways, moving swiftly; or data of hosts of angels, not absolute, or still dynamic, I suspect, myself, that the fixed stars are really fixed, and that the minute motions said to have been detected in them are illusions. I think that the fixed stars are absolutes. Their twinkling is only the interpretation by an intermediatist state of them. I think that soon after Leverrier died, a new fixed star was discovered—that, if Dr. Gray had stuck to his story of the thousands of fishes from one pail of water, had written upon it, taken to street corners, to convince the world that, whether conceivable or not, his explanation was the only true explanation: had thought of nothing but this last thing at night and first thing in the morning—his obituary—another "nova" reported in *Monthly Notices*.

I think that Milky Ways, of an inferior, or dynamic, order, have often been seen by astronomers. Of course it may be that the phenomena that we shall now consider are not angels at all. We are simply feeling around, trying to find out what we can accept. Some of our data indicate hosts of rotund and complacent tourists in inter-planetary space—but then data of long, lean, hungry ones. I think that there are, out in inter-planetary space, Super Tamerlanes at the head of hosts of celestial ravagers—which have come here and pounced upon civilizations of the past, cleaning them up all but their bones, or temples and monuments—for which later historians have invented exclusionist histories. But if something now has a legal right to us, and can enforce its proprietorship, they've been warned off. It's the way of all exploitation. I should say that we're now under cultivation: that we're conscious of it, but have the impertinence to attribute it all to our own nobler and higher instincts.

Against these notions is the same sense of finality that opposes all advance. It's why we rate acceptance as a better adaptation than belief. Opposing us is the strong belief that, as to inter-planetary phenomena, virtually everything has been found out. Sense of finality and illusion of homogeneity. But that what is called advancing knowledge is violation of the sense of blankness.

A drop of water. Once upon a time water was considered so homogeneous that it was thought of as an element. The microscope—and not only that the supposititiously elementary was seen to be of infinite diversity, but that in his protoplasmic life there were new orders of beings.

Or the year 1491—and a European looking westward over the ocean—his feeling that that suave western droop was unbreakable; that gods of regularity would not permit that smooth horizon to be disturbed by coasts or spotted with islands. The unpleasantness of even contemplating such a state—wide, smooth west, so clean against the sky—spotted with islands—geographic leprosy.

But coasts and islands and Indians and bison, in the seemingly vacant west: lakes, mountains, rivers—

One looks up at the sky: the relative homogeneity of the relatively unexplored: one thinks of only a few kinds of phenomena. But the acceptance is forced upon me that there are modes and modes and modes of inter-planetary existence: things as different from planets and comets and meteors as Indians are from bison and prairie dogs: a super-geography—or celestiography—of vast stagnant regions, but also of Super-Niagaras and Ultra-Mississippis: and a super-sociology—voyagers and tourists and ravagers: the hunted and the hunting: the super-mercantile, the super-piratic, the super-evangelical.

Sense of homogeneity, or our positivist illusion of the unknown—and the fate of all positivism.

Atronomy and the academic.

Ethics and the abstract,

The universal attempt to formulate or to regularize—an attempt that can be made only by disregarding or denying.

Or all things disregard or deny that which will eventually invade and destroy them—

227

Until comes the day when some one thing shall say, and enforce upon Infinitude:

"Thus far shalt thou go: here is absolute demarcation."

The final utterance:

"There is only I."

In the *Monthly Notices of the R.A.S.*, 11–48, there is a letter from the Rev. W. Read:

That, upon the 4th of September, 1851, at 9.30 A.M., he had seen a host of self-luminous bodies, passing the field of his telescope, some slowly and some rapidly. They appeared to occupy a zone several degrees in breadth. The direction of most of them was due east to west, but some moved from north to south. The numbers were tremendous. They were observed for six hours.

Editor's note:

"May not these appearances be attributed to an abnormal state of the optic nerves of the observer?"

In *Monthly Notices*, 12–38, Mr. Read answers that he had been a diligent observer, with instruments of a superior order, for about 28 years—"but I have never witnessed such an appearance before". As to illusion he says that two other members of his family had seen the objects.

The Editor withdraws his suggestion.

We know what to expect. Almost absolutely—in an existence that is essentially Hibernian—we can predict the past—that is, look over something of this kind, written in 1851, and know what to expect from the Exclusionists later. If Mr. Read saw a migration of dissatisfied angels, numbering millions, they must merge away, at least subjectively with common-place terrestrial phenomena—of course disregarding Mr. Read's probable familiarity, of 28 years' duration, with the commonplaces of terrestrial phenomena.

Monthly Notices, 12–183:

Letter from Rev. W. R. Dawes:

That he had seen similar objects—and in the month of September—that they were nothing but seeds floating in the air.

In the *Report of the British Association*, 1852–235, there is a communication from Mr. Read to Prof. Baden Powell:

That the objects that had been seen by him and by Mr. Dawes were not similar. He denies that he had seen seeds floating in the air. There had been little wind, and that had come from the sea, where seeds would not be likely to have origin. The objects that he had seen were round and sharply defined, and with none of the feathery appearance of thistledown. He then quotes from a letter, from C. B. Chalmers, F.R.A.S., who had seen a similar stream, a procession, or migration, except that some of the bodies were more elongated—or lean and hungry—than globular.

He might have argued for sixty-five years. He'd have impressed nobody—of importance. The super-motif, or dominant, of his era, was Exclusionism, and the notion of seeds in the air assimilates—with due disregards—with that dominant.

Or pageantries here upon our earth, and things looking down upon us—and the Crusades were only dust clouds, and glints of the sun on shining armour were only particles of mica in dust clouds. I think it was a Crusade that Read saw—but that it was right, relatively to the year 1851, to say that it was only seeds in the wind, whether the wind blew from the sea or not. I think of things that were luminous with religious zeal, mixed up, like everything else in Intermediateness, with black marauders and from grey to brown beings of little personal ambitions. There may have been a Richard Coeur de Lion, on his way to right wrongs in Jupiter. It was right, relatively to 1851, to say that he was a seed of a cabbage.

Prof. Coffin, U.S.N. (*Jour. Frank. Inst.*, 88–151):

That, during the eclipse of August, 1869, he had noted the passage, across his telescope, of several bright flakes resembling thistleblows, floating in the sunlight. But the telescope was so focused that, if these things were distinct, they must have been so far away from this earth that the difficulties of orthodoxy remain as great, one way or another, no matter what we think they were—

They were "well-defined", says Prof. Coffin.

Henry Waldner (*Nature*, 5–304):

That, April 27, 1863, he had seen great number of small shining bodies passing from west to east. He had notified Dr. Wolf, of the Observatory of Zurich, who "had convinced himself of this strange phenomenon". Dr. Wolf had told him that similar bodies had been seen by Sig. Capocci, of the Capodimonte Observatory at Naples, May 11, 1845.

The shapes were of great diversity—or different aspects of similar shapes?

Appendages were set upon some of them.

We are told that some were star-shaped, with transparent appendages.

I think, myself, it was a Mohammed and his Hegira. May have been only his harem. Astonishing sensation: afloat in space with ten million wives around one. Anyway, it would seem that we have considerable advantage here, inasmuch as seeds are not in season in April—but the pulling back to earth, the bedraggling by those sincere but dull ones of some time ago. We have the same stupidity—necessary, functioning stupidity—of attribution of something that was so rare that an astronomer notes only one instance between 1845 and 1863, to an everyday occurrence—

Or Mr. Waldner's assimilative opinion that he had seen only ice crystals.

Whether they were not very exclusive veils of a super-harem, or planes of a very light material, we have an impression of star-shaped things with transparent appendages that have been seen in the sky.

Hosts of small bodies—black, this time—that were seen by the astronomers Herrick, Buys-Ballot, and De Cuppis (*L'Année Scientifique*, 1860–25); vast numbers of bodies that were seen by M. Lamey, to cross the moon (*L'Année Scientifique*, 1874–62); another instance of dark ones; prodigious number of dark, spherical bodies reported by Messier, June 17, 1777 (Arago, *Oeuvres* 9–38); considerable number of luminous bodies which appeared to move out from the sun, in diverse directions; seen at Havana, during eclipse of the sun, May 15, 1836, by Prof. Auber (Poey); M. Poey cites a similar instance, of Aug. 3, 1886; M. Lotard's opinion that they were birds (*L'Astronomie*, 1886–391); large number of small bodies crossing disc of the sun, some swiftly, some slowly, most of them globular, but some seemingly triangular, and some of more complicated structure; seen by M. Trouvelet, who, whether seeds, insects, birds, or other commonplace things had never seen anything resembling these forms (*L'Année Scientifique*, 1885–8); report from the Rio de Janeiro Observatory, of vast numbers of bodies crossing the sun, some of them luminous and some of them dark, from some time in December, 1875, until Jan. 22, 1876 (*La Nature*, 1876–384).

Of course, at a distance, any form is likely to look round or roundish: but we point out that we have notes upon the seeming of more complex forms. In *L'Astronomie*, 1886–70, is recorded M. Briguiere's observation at Marseilles, April 15 and April 25, 1883, upon the crossing of the sun by bodies that were irregular in form. Some of them moved as if in alignment.

Letter from Sir Robert Inglis to Col. Sabine (*Rept. Brit. Assoc.*, 1849–17):

That, at 3 P.M., at Gais, Switzerland, Inglis had seen thousands and thousands of brilliant white objects, like snowflakes in a cloudless sky. Though this display lasted about twenty-five minutes, not one of these seeming snowflakes was seen to fall. Inglis says that his servant "fancied" that he had seen something like wings on these—whatever they were. Upon page 18, of the Report, Sir John Herschel says that, in 1845 or 1846, his attention had been attracted by objects of considerable size, in the air, seemingly not far away. He says that they were masses of hay, not less than a yard or two in diameter. Still there are some circumstances that interest me. He says that, though no less than a whirlwind could have sustained these masses, the air about was calm. "No doubt wind prevailed at the spot, but there was no roaring noise." None of these masses fell within his observation or knowledge. To walk a few fields away and find out more would seem not much to expect from a man of science, but it is one of our superstitions, that such a seeming trifle is just what—by the Spirit of an Era, we'll call it—one is not permitted to do. If those things were not masses of hay, and if Herschel had walked a little and found out, and had reported that he had seen strange objects in the air— that report, in 1846, would still have been misplaced as the appearance of a tail upon an embryo still in its gastrula era. I have noticed this inhibition in my own case many times. Looking back—why didn't I do this or that little thing that would have cost so little and have meant so much? Didn't belong to that era of my own development.

Nature, 22–64:

That, at Kattenau, Germany, about half an hour before sunrise, March 22, 1880, "an enormous number of luminous bodies rose from the horizon, and passed in a horizontal

231

direction from east to west". They are described as having appeared in a zone or belt. "They shone with a remarkably brilliant light."

So they've thrown lassos over our data to bring them back to earth. But they're lassos that cannot tighten. We can't pull out of them: we may step out of them, or lift them off. Some of us used to have an impression of Science sitting in calm, just judgment: some of us now feel that a good many of our data have been lynched. If a Crusade, perhaps from Mars to Jupiter, occur in the autumn—"seeds". If a Crusade or outpouring of celestial vandals is seen from this earth in the spring—"ice crystals". If we have record of a race of aerial beings, perhaps with no substantial habitat, seen by someone in India—"locusts".

This will be disregarded:

If locusts fly high, they freeze and fall in thousands.

Nature, 47–581:

Locusts that were seen in the mountains of India, at a height of 12,750 feet—"in swarms and dying by thousands".

But no matter whether they fly high or low, no one ever wonders what's in the air when locusts are passing overhead, because of the falling of stragglers. I have especially looked this matter up—no mystery when locusts are flying overhead—constant falling of stragglers.

Monthly Notices, 30–135:

"An unusual phenomenon noticed by Lieut. Herschel, Oct. 17 and 18, 1870, while observing the sun, at Bangalore, India."

Lieut. Herschel had noticed dark shadows crossing the sun—but away from the sun there were luminous, moving images. For two days bodies passed in a continuous stream, varying in size and velocity.

The Lieutenant tries to explain, as we shall see, but he says:

"As it was, the continuous flight, for two whole days, in such numbers in the upper regions of the air, of beasts that left no stragglers, is a wonder of natural history, if not of astronomy."

He tried different focusing—he saw wings—perhaps he saw planes. He says that he saw upon the objects either wings or phantom-like appendages.

Then he saw something that was so bizarre that, in the

fullness of his nineteenth-centuriness, he writes:

"There was no longer doubt: they were locusts or flies of some sort."

One of them had paused.

It had hovered.

Then it had whisked off.

The Editor says that at that time "countless locusts had descended upon certain parts of India".

We now have an instance that is extraordinary in several respects—super-voyagers or super-ravagers; angels, ragamuffins, crusaders, emigrants, aeronauts, or aerial elephants, or bison or dinosaurs—except that I think the thing had planes or wings—one of them has been photographed. It may be that in the history of photography no more extraordinary picture than this has ever been taken.

L'Astronomie, 1885–347:

That, at the Observatory of Zacatecas, Mexico, Aug. 12, 1883, about 2,500 metres above sea level, were seen a large number of small luminous bodies, entering upon the disc of the sun. M. Bonilla telegraphed to the Observatories of the City of Mexico and of Puebla. Word came back that the bodies were not visible there. Because of this parallax, M. Bonilla placed the bodies "relatively near the earth". But when we find out what he called "relatively near the earth"—birds or bugs or hosts of a Super-Tamerlane or army of a celestial Richard Coeur de Lion—our heresies rejoice anyway. His estimate is "less distance than the moon".

One of them was photographed. See *L'Astronomie*, 1885–349. The photographs shows a long body surrounded by indefinite structures, or by the haze of wings or planes in motion.

L'Astronomie, 1887–66:

Signor Ricco, of the Observatory of Palermo, writes that, Nov. 30, 1880, at 8.30 o'clock in the morning, he was watching the sun, when he saw, slowly traversing its disc, bodies in two long parallel lines, and a shorter, parallel line. The bodies looked winged to him. But so large were they that he had to think of large birds. He thought of cranes.

He consulted ornithologists, and learned that the configuration of parallel lines agrees with the flight-formation of cranes. This was in 1880: anybody now living in New

York City, for instance, would tell him that also it is a familiar formation of aeroplanes. But, because of data of focus and subtended angles, these beings or objects must have been high.

Our own acceptance, in conventional terms, is that there is not a bird of this earth that would not freeze to death at a height of more than four miles: that if condors fly three or four miles high, they are birds that are especially adapted to such altitudes.

Sig. Ricco's estimate is that these objects or beings or cranes must have been at least five and a half miles high.

Chapter 17

The vast dark thing that looked like a poised crow of unholy
dimensions. Assuming that I shall ever have any readers, let
him, or both of them, if I shall ever have such popularity as
that, note how dim that bold black datum is at the distance
of only two chapters.

The question:

Was it a thing or the shadow of a thing?

Acceptance either way calls not for mere revision but
revolution in the science of astronomy. But the dimness of
the datum of only two chapters ago. The carved stone disc
of Tarbes, and the rain that fell every afternoon for
twenty—if I haven't forgotten, myself, whether it was
twenty-three or twenty-five days—upon one small area.
We are all Thomsons, with brains that have smooth and
slippery, though corrugated, surfaces—or that all intellection is associative—or that we remember that which correlates with a dominant—and a few chapters go by, and
there's scarcely an impression that hasn't slid off our
smooth and slippery brains, of Leverrier and the "planet
Vulcan". There are two ways by which irreconcilables can
be remembered—if they can be correlated in a system
more nearly real than the system that rejects them—and by
repetition and repetition and repetition.

Vast black thing like a crow poised over the moon.

The datum is so important to us, because it enforces, in
another field, our acceptance that dark bodies of planetary
size traverse this solar system.

Our position:

That the things have been seen:

Also that their shadows have been seen.

Vast black thing poised like a crow over the moon. So far
it is a single instance. By a single instance, we mean the
negligible.

In *Popular Science*, 34–158, Serviss tells of a shadow that
Schroeter saw, in 1788, in the lunar Alps. First he saw a
light. But then, when the region was illuminated, he saw a
round shadow where the light had been.

Our own expression:

That he saw a luminous object near the moon: that that part of the moon became illuminated, and the object was lost to view; but that then its shadow underneath was seen.

Serviss explains, of course. Otherwise he'd not be Prof. Serviss. It's a little contest in relative approximations to realness. Prof. Serviss thinks that what Schroeter saw was the "round" shadow of a mountain—in the region that had become lighted. He assumes that Schroeter never looked again to see whether the shadow could be attributed to a mountain. That's the crux: conceivably a mountain could cast a round—and that means detached—shadow, in the lighted part of the moon. Prof. Serviss could, of course, explain why he disregards the light in the first place—maybe it has always been there "in the first place". If he couldn't explain, he'd still be an amateur.

We have another datum. I think it is more extraordinary than—

Vast thing, black and poised, like a crow, over the moon.

But only because it's more circumstantial, and because it has corroboration, do I think it more extraordinary than—

Vast poised thing, black as a crow, over the moon.

Mr. H. C. Russell, who was usually as orthodox as anybody, I suppose—at least, he wrote "F.R.A.S." after his name—tells in the *Observatory*, 2–374, one of the wickedest, or most preposterous, stories that we have so far exhumed:

That he and another astronomer, G. D. Hirst, were in the Blue Mountains, near Sydney, N.S.W., and Mr. Hirst was looking at the moon—

He saw on the moon what Russell calls "one of those remarkable facts, which being seen should be recorded, although no explanation can at present be offered".

That may be so. It is very rarely done. Our own expression upon evolution by successive dominants and their correlates is against it. On the other hand, we express that every era records a few observations out of harmony with it, but adumbratory or preparatory to the spirit of eras still to come. It's very rarely done. Lashed by the phantom-scourge of a now passing era, the world of astronomers is in a state of terrorism, though of a highly attenuated, modernized, devitalized kind. Let an astronomer see something that is not of the conventional, celestial sights, or something that it is "improper" to see—his very dignity is in danger. Some one of the corralled and scourged may stick a smile into his back. He'll be thought of unkindly.

With a hardihood that is unusual in his world of ethereal sensitivenesses, Russell says, of Hirst's observation:

"He found a large part of it covered with a dark shade, quite as dark as the shadow of the earth during an eclipse of the moon."

But the climax of hardihood or impropriety or wickedness, preposterousness or enlightenment:

"One could hardly resist the conviction that it was a shadow, yet it could not be the shadow of any known body."

Richard Proctor was a poor man of some liberality. After a while we shall have a letter, which once upon a time we'd have called delirious—don't know that we could read such a thing now, for the first time, without incredulous laughter—which Mr. Proctor permitted to be published in *Knowledge*. But a dark, unknown world that could cast a shadow upon a large part of the moon, perhaps extending far beyond the limb of the moon; a shadow as deep as the shadow of this earth—

Too much for Mr. Proctor's politeness.

I haven't read what he said, but it seems to have been a little coarse. Russell says that Proctor "freely used" his name in the *Echo*, of March 14, 1879, ridiculing this observation which had been made by Russell as well as Hirst. If it hadn't been Proctor, it would have been someone else—but one notes that the attack came out in a newspaper. There is no discussion of this remarkable subject, no mention in any were open to Russell to answer Proctor.

In the answer, I note considerable intermediateness. Far back in 1879, it would have been a beautiful positivism, if Russell had said—

"There was a shadow on the moon. Absolutely it was cast by an unknown body."

According to our religion, if he had then given all his time to the maintaining of this one stand, of course breaking all friendships, all ties with his fellow astronomers, his apotheosis would have occurred, greatly assisted by means well known to quasi-existence when its compromises and evasions, and phenomena that are partly this and partly that, are flouted by the definite and uncompromising. It would be impossible in a real existence, but Mr. Russell, of quasi-existence, says that he did resist the conviction; that he had said that one could "hardly resist"; and most of his resentment is against Mr. Proctor's thinking that he had not resisted. It seems too bad—if apotheosis be desirable.

The point in Intermediatism here is:

Not that to adapt to the conditions of quasi-existence is to have what is called success in quasi-existence, but is to lose one's soul—

But is to lose "one's" chance of attaining soul, self, or entity.

One indignation quoted from Proctor interests us:

"What happens on the moon may at any time happen to this earth."

Or:

That is just the teaching of this department of Advanced Astronomy:

That Russell and Hirst saw the sun eclipsed relatively to the moon by a vast dark body;

That many times have eclipses occurred relatively to this earth, by vast, dark bodies;

That there have been many eclipses that have not been recognized as eclipses by scientific kindergartens.

There is a merger, of course. We'll take a look at it first—that, after all, it may have been a shadow that Hirst and Russell saw, but the only significance is that the sun was eclipsed relatively to the moon by a cosmic haze of some kind, or a swarm of meteors close together, or a gaseous discharge left behind by a comet. My own acceptance is that vagueness of shadow is a function of vagueness of intervention; that a shadow as dense as the shadow of the earth is cast by a body denser than hazes and swarms. The information seems definite enough in this respect—"Quite as dark as the shadow of this earth during the eclipse of the moon".

Though we may not always be as patient toward them as we should be, it is our acceptance that the astronomic primitives have done a great deal of good work: for instance, in the allaying of fears upon this earth. Sometimes it may seem as if all science were to us very much like what unsquare meals are to bulls and anti-socialists—not the scientific, but the insufficient. Our acceptance is that Evil is the negative state, by which we mean the state of maladjustment, discord, ugliness, disorganization, inconsistency,

238

injustice, and so on—as determined in Intermediateness, not by real standards, but only by higher approximations to adjustment, harmony, beauty, organization, consistency, justice, and so on. Evil is outlived virtue or incipient virtue that has not yet established itself or any other phenomenon that is not in seeming adjustment, harmony, consistency with a dominant. The astronomers have functioned bravely in the past. They've been good for business: the big interests think kindly, if at all, of them. It's bad for trade to have an intense darkness come upon an unaware community and frighten people out of their purchasing values. But if an obscuration be foretold, and if it then occur—may seem a little uncanny—only a shadow—and no one who was about to buy a pair of shoes runs home panic-stricken and saves the money.

Upon general principles we accept that astronomers have quasi-systematized data of eclipses—or have included some and disregarded others.

They have done well.

They have functioned.

But now they're negatives, or they're out of harmony—

If we are in harmony with a new dominant, or the spirit of a new era, in which Exclusionism must be overthrown; if we have data of many obscurations that have occurred, not only upon the moon, but upon our own earth, as convincing of vast intervening bodies, usually invisible, as is any regularized, predicted eclipse.

One looks up at the sky.

It seems incredible that, say, at the distance of the moon, there could be, but be invisible, a solid body, say, the size of the moon.

One looks up at the moon at a time when only a crescent of it is visible. The tendency is to build up the rest of it in one's mind; but the unillumined part looks as vacant as the rest of the sky, and it's of the same blueness as the rest of the sky. There's a vast area of solid substance before one's eyes. It's indistinguishable from the sky.

In some of our little lessons upon the beauties of modesty and humility, we have picked out basic arrogances—tail of a peacock, horns of a stag, dollars of a capitalist—eclipses of astronomers. Though I have no desire for the job, I'd engage to list hundreds of instances in which the report upon an expected eclipse has been "sky overcast" or "weather unfavourable". In our Super-Hibernia, the unfavourable has been construed as the favourable. Some time ago, when we were lost, because we had not recognized our own dominant, when we were still of the unchosen and likely to be more malicious than we now are—because we have noted a steady tolerance creeping into our attitude—if astronomers are not to blame, but are only correlates to a dominant—we advertised a predicted eclipse that did not occur at all. Now, without any especial feeling, except that of recognition of the fate of all attempted absolutism, we give the instance, noting that, though such an evil thing to orthodoxy, it was orthodoxy that recorded the non-event.

Monthly Notices of the R.A.S., 8–132:

"Remarkable appearances during the total eclipse of the moon on March 19, 1848":

In an extract from a letter from Mr. Forester, of Bruges, it is said that, according to the writer's observations at the time of the predicted total eclipse, the moon shone with about three times the intensity of the mean illumination of an eclipsed lunar disc: that the British Consul, at Ghent, who did not know of the predicted eclipse, had written inquiring as to the "blood-red" colour of the moon.

This is not very satisfactory to what used to be like our malices. But there follows another letter, from another astronomer, Walkey, who had made observations at Clyst St. Lawrence: that, instead of an eclipse, the moon became—as is printed in italics—"most beautifully illuminated" ... "rather tinged with a deep red" ... "the moon being as perfect with light as if there had been no eclipses whatever".

I note that Chambers, in his work upon eclipses, gives Forster's letter in full—and not a mention of Walkey's letter.

There is no attempt in *Monthly Notices* to explain upon the notion of greater distance of the moon, and the earth's shadow falling short, which would make as much trouble for astronomers, if that were not foreseen, as no eclipse at all. Also there is no refuge in saying that virtually never, even in total eclipses, is the moon totally dark—"as perfect with light as if there had been no eclipse whatever". It is said that at the time there had been an aurora borealis, which might have caused the luminosity, without a datum that such an effect, by an aurora, had ever been observed upon the moon.

But single instances—so an observation by Scott, in the Antarctic. The force of this datum lies in my own acceptance, based upon especially looking up this point, that an eclipse nine-tenths of totality has a greater effect, even though the sky be clouded.

Scott (*Voyage of Discovery*, vol. II, p. 215):

"There may have been an eclipse of the sun, Sept. 21, 1903, as the almanac said, but we should, none of us, have liked to swear to the fact."

This eclipse had been set down at nine-tenths of totality. The sky was overcast at the time.

So it is not only that many eclipses unrecognized by astronomers as eclipses have occurred, but that intermediatism, or impositivism, breaks into their own seemingly regularized eclipses.

Our data of unregularized eclipses, as profound as those that are conventionally—or officially?—recognized, that have occurred relatively to this earth:

In *Notes and Queries* there are several allusions to intense darknesses that have occurred upon this earth, quite as eclipses occur, but that are not referable to any known eclipsing body. Of course there is no suggestion here that these darknesses may have been eclipses. My own acceptance is that if in the nineteenth century anyone had uttered such a thought as that, he'd have felt the blight of a Dominant; that Materialistic Science was a jealous god, excluding, as works of the devil, all utterances against the seemingly uniform, regular periodic; that to defy him would have brought on—withering by ridicule—shrinking away by publishers—contempt of friends and family—justifiable grounds for divorce—that one who would so defy would feel what unbelievers in relics of saints felt in an earlier age; what befell virgins who forgot to keep fires burning, in a still earlier age—but that, if he'd almost absolutely hold out, just the same—new fixed star reported in *Monthly Notices*. Altogether, the point in Positivism here is that by Dominants and their correlates, quasi-existence strives for the positive state, aggregating, around a nucleus, or dominant, systematized members of a religion, a science, a society—but that "individuals" who do not surrender and submerge may of themselves highly approximate to

241

positiveness—the fixed, the real, the absolute.

In *Notes and Queries*, 2–4–139, there is an account of a darkness in Holland, in the midst of a bright day, so intense and terrifying that many panic-stricken persons lost their lives stumbling into the canals.

Gentleman's Magazine, 33–414:

A darkness that came upon London, Aug. 19, 1763, "greater than at the great eclipse of 1748".

However, our preference is not to go so far back for data. For a list of historic "dark days", see Humboldt, *Cosmos*, 1–120.

Monthly Weather Review, March, 1886–79:

That, according to the *La Cross Daily Republican*, of March 20, 1886, darkness suddenly settled upon the city of Oshkosh, Wis., at 3 P.M., March 19. In five minutes the darkness equalled that of midnight.

Consternation.

I think that some of us are likely to overdo our own superiority and the absurd fears of the Middle Ages—

Oshkosh.

People in the streets rushing in all directions—horses running away—women and children running into cellars—little modern touch after all: gas meters instead of images and relics of saints.

This darkness, which lasted for eight to ten minutes, occurred in a day that had been "light but cloudy". It passed from west to east, and brightness followed: then came reports from towns to the west of Oshkosh: that the same phenomenon had already occurred there. A "wave of total darkness" had passed from west to east.

Other instances are recorded in the *Monthly Weather Review*, but, as to all of them, we have a sense of being pretty well-eclipsed, ourselves, by the conventional explanation that the obscuring body was on a very dense mass of clouds. But some of the instances are interesting—intense darkness at Memphis, Tenn., for about fifteen minutes, at 10 A.M., Dec. 2, 1904—"We are told that in some quarters a panic prevailed, and that some were shouting and praying and imagining that the end of the world had come." (*M.W.R.*, 32–522.) At Louisville, Ky., March 7, 1911, at about 8 A.M.: duration about half an hour; had been raining moderately, and then hail had fallen. "The intense blackness and general ominous

appearance of the storm spread terror throughout the city."
(*M.W.R.*, 39–345.)

However, this merger between possible eclipses by unknown dark bodies and commonplace terrestrial phenomena is formidable.

As to darknesses that have fallen upon vast areas, conventionality is—smoke from forest fires. In the *U.S. Forest Service Bulletin*, No. 117, F. G. Plummer gives a list of eighteen darknesses that have occurred in the United States and Canada. He is one of the primitives, but I should say that his dogmatism is shaken by vibrations from the new Dominant. His difficulty, which he acknowledges, but which he would have disregarded had he written a decade or so earlier, is the profundity of some of these obscurations. He says that mere smokiness cannot account for such "awe inspiring dark days". So he conceives of eddies in air, concentrating the smoke from forest fires. Then, in the inconsistency or discord of all quasi-intellection that is striving for consistency or harmony, he tells of the vastness of some of these darknesses. Of course Mr. Plummer did not really think upon this subject, but one does feel that he might have approximated higher to real thinking than by speaking of concentration and then listing data of enormous area, or the opposite of circumstances of concentration—because, of his nineteen instances, nine are set down as covering all New England. In quasi-existence, everything generates or is part of its own opposite. Every attempt at peace prepares the way for war; all attempts at justice result in injustice in some other respect: so Mr. Plummer's attempt to bring order into his data, with the explanation of darkness caused by smoke from forest fires, results in such confusion that he ends up by saying that these daytime darknesses have occurred "often with little or no turbidity of the air near the earth's surface"—or with no evidence at all of smoke—except that there is almost always a forest fire somewhere.

However, of the eighteen instances, the only one that I'd bother to contest is the profound darkness in Canada and northern parts of the United States, Nov. 19, 1819—which we have already considered.

Its concomitants:

Lights in the sky;

Fall of a black substance;

Shocks like those of an earthquake.

In this instance, the only available forest fire was one to the south of the Ohio River. For all I know, soot from a very great fire south of the Ohio River might fall in Montreal, Canada, and conceivably, by some freak of reflection, light from it might be seen in Montreal, but the earthquake is not assimilable with a forest fire. On the other hand, it will soon be our expression that profound darkness, fall of matter from the sky, lights in the sky, and earthquakes are phenomena of the near approach of other worlds to this world. It is such comprehensiveness, as contrasted with inclusion of a few factors and disregard for the rest, that we call higher approximation to realness—or universalness.

A darkness, of April 17, 1904, at Wimbledon, England (*Symons' Met. Mag.*, 39–69). It came from a smokeless region: no rain, no thunder; lasted 10 minutes; too dark to go "even out in the open".

As to darknesses in Great Britain, one thinks of fogs —but in *Nature*, 25–289, there are some observations by Major J. Herschel, upon an obscuration in London, Jan. 22, 1882, at 10.30 A.M., so great that he could hear persons upon the opposite side of the street, but could not see them—"It was obvious that there was no fog to speak of."

Annual Register, 1857—132:

An account by Charles A. Murray, British Envoy to Persia, of a darkness of May 20, 1857, that came upon Baghdad—"a darkness more intense than ordinary midnight, when neither stars nor moon are visible...." "After a short time the black darkness was succeeded by a red, lurid gloom, such as I never saw in any part of the world."

"Panic seized the whole city."

"A dense volume of red sand fell."

This matter of sand falling seems to suggest conventional explanation enough, or that a simoon, heavily charged with terrestrial sand had obscured the sun, but Mr. Murray, who says that he had had experience with simoons, gives his opinion that "it cannot have been a simoon".

It is our comprehensiveness now, or this matter of concomitants of darknesses that we are going to capitalize. It is all very complicated and tremendous, and our own treatment can be but impressionistic, but a few of the rudiments of Advanced Seismology we shall now take up—or the four principal phenomena of another world's close approach to this world.

If a large substantial mass, or super-construction, should enter this earth's atmosphere, it is our acceptance that it would sometimes—depending upon velocity—appear luminous or look like a cloud, or look like a cloud with a luminous nucleus. Later we shall have an expression upon luminosity—different from the luminosity of incandescence—that comes upon objects falling from the sky, or entering this earth's atmosphere. Now our expression is that worlds have often come close to this earth, and that smaller objects—size of a haystack or size of several dozen skyscrapers lumped, have often hurtled through this earth's atmosphere, and have been mistaken for clouds, because they were enveloped in clouds—

Or that around something coming from the intense cold of inter-planetary space—that is of some regions: our own suspicion is that other regions are tropical—the moisture of this earth's atmosphere would condense into a cloud-like appearance around it. In *Nature*, 20-121, there is an account by Mr. S. W. Clifton, Collector of Customs, at Freemantle, Western Australia, sent to the Melbourne Observatory—a clear day—appearance of a small black cloud, moving not very swiftly—bursting into a ball of fire, of the apparent size of the moon—

Or that something with the velocity of an ordinary meteorite could not collect vapour around it, but that slower-moving objects—speed of a railway train, say—may.

The clouds of tornadoes have so often been described as if they were solid objects that I now accept that sometimes they are: that some so-called tornadoes are objects hurtling through this earth's atmosphere, not only generating disturbances by their suctions, but crushing, with their bulk, all things in their way, rising and falling and finally disappearing, demonstrating that gravitation is not the power that the primitives think it is, if an object moving at relatively low velocity be not pulled to this earth, or being so momentarily affected, bounds away.

Or that it was no meteorological phenomenon, but something very suggestive bits of description occur:

"Cloud bounded along the earth like a ball"—

In Finley's *Reports on the Character of 600 Tornadoes* very much like a huge solid ball that was bounding along, crushing and carrying with it everything within its field—

"Cloud bounded along, coming to the earth every eight hundred or one thousand yards."

Here's an interesting bit that I got somewhere else. I offer it as a datum in super-biology, which, however, is a branch of advanced science that I'll not take up, restricting to things indefinitely called "objects"—

"The tornado came wriggling, jumping, whirling like a great green snake, darting out a score of glistening fangs."

Though it's interesting, I think that's sensational, myself. It may be that vast green snakes sometimes rush past this earth, taking a swift bite wherever they can, but, as I say, that's a super-biologic phenomenon. Finley gives dozens of instances of tornado clouds that seem to me more like solid things swathed in clouds, than clouds. He notes that, in the tornado at Americus, Georgia, July 18, 1881, "a strange sulphurous vapour was emitted from the cloud". In many instances, objects, or meteoritic stones, that have come from this earth's externality have had a sulphurous odour. Why a wind effect should be sulphurous is not clear. That a vast object from external regions should be sulphurous is in line with many data. This phenomenon is described in the *Monthly Weather Review*, July, 1881, as "a strange sulphurous vapour ... burning and sickening all who approached close enough to breathe it".

The conventional explanation of tornadoes as wind-effects—which we do not deny in some instances—is so strong in the United States that it is better to look elsewhere for an account of an object that has hurtled through this earth's atmosphere, rising and falling and defying this earth's gravitation.

Nature, 7–112:

That, according to a correspondent to the *Birmingham Morning News*, the people living near King's Sutton, Banbury, saw, about one o'clock, Dec. 7, 1872, something like a haycock hurtling through the air. Like a meteor it was accompanied by fire and dense smoke and made a noise like that of a railway train. "It was sometimes high in the air and sometimes near the ground." The effect was tornado-like: trees and walls were knocked down. It's a late day now to try to verify this story, but a list is given of persons whose property was injured. We are told that this thing then disappeared "all at once".

These are the smaller objects, which may be derailed railway trains or big green snakes, for all I know—but our expression upon approach to this earth by vast dark bodies—

That likely they'd be made luminous: would envelop in clouds, perhaps, or would have their own clouds—

But that they'd quake, and that they'd affect this earth with quakes—

And that then would occur a fall of matter from such a world, or rise of matter from this earth to a nearby world, or both fall and rise, or exchange of matter—process known to Advanced Seismology as celestio-metathesis—

Except that—if matter from some other world—and it would be like someone to get it into his head that we absolutely deny gravitation, just because we cannot accept orthodox dogmas—except that, if matter from another world, filling the sky of this earth, generally, as to a hemisphere, or locally, should be attracted to this earth, it would seem thinkable that the whole thing should drop here and not merely its surface-materials.

Objects upon a ship's bottom. From time to time they drop to the bottom of the ocean. The ship does not.

Or, like our acceptance upon dripping from aerial ice-fields, we think of only a part of a nearby world succumbing, except in being caught in suspension, to this earth's gravitation, and surface-materials falling from that part—

Explain or express or accept, and what does it matter? Our attitude is:

Here are the data.

See for yourself.

What does it matter what my notions may be?

Here are the data.

But think for yourself, or think for myself, all mixed up we must be. A long time must go by before we can know Florida from Long Island. So we've had data of fishes that have fallen from our now established and respectabilized Super-Sargasso Sea—which we've almost forgotten, it's now so respectable—but we shall have data of fishes that have fallen during earthquakes. These we accept were dragged from ponds or other worlds that have been quaked, when only a few miles away, by this earth, some other world also quaking this earth.

In a way, or in its principle, our subject is orthodox

enough. Only grant proximity of other worlds—which, however, will not be a matter of granting, but will be a matter of data—and one conventionally conceives of their surfaces quaked—even of a whole lake full of fishes being quaked and dragged from one of them. The lake full of fishes may cause a little pain to some minds, but the fall of sand and stones is pleasantly enough thought of. More scientific persons, or more faithful hypnotics than we, have taken up this subject, unpainfully, relatively to the moon. For instance, Perry has gone over 15,000 records of earthquakes, and he has correlated many with proximities of the moon or has attributed many to the pull of the moon when nearest to the earth. Also there is a paper upon this subject in the *Proc. Roy. Soc. of Cornwall*, 1845. Or, theoretically, when at its closest to this earth, the moon quakes the face of this earth, and is itself quaked—but does not itself fall to this earth. As to showers of matter that may have come from the moon at such times—one can go over old records and find what one pleases.

That is what we now shall do.

Our expressions are for acceptance only.

Our data:

We take them from four classes of phenomena that have preceded or accompanied earthquakes:

Unusual clouds, darkness profound, luminous appearances in the sky, and falls of substances and objects whether commonly called meteoritic or not.

Not one of these occurrences fits in with principles of primitive, or primary, seismology, and every one of them is a datum of a quaked body passing close to this earth or suspended over it. To the primitives there is not a reason in the world why a convulsion of this earth's surface should be accompanied by unusual sights in the sky, by darkness, or by the fall of substances or objects from the sky. As to phenomena like these, or storms, preceding earthquakes, the irreconcilability is still greater.

It was before 1860 that Perrey made his great compilation. We take most of our data from lists compiled long ago. Only the safe and unpainful have been published in recent years—at least in ambitious, voluminous form. The restraining hand of the "System"—as we call it, whether it has any real existence or not—is tight upon the sciences of today. The uncanniest aspect of our quasi-experience that I know of is that everything that seems to have one identity has also as high a seeming of everything else. In this oneness of allness, or continuity, the protecting hand strangles; the parental stifles; love is inseparable from phenomena of hate. There is only Continuity—that is in quasi-existence. *Nature*, at least in its correspondents' columns, still evades this protective strangulation, and the *Monthly Weather Review* is still a rich field of unfaithful observation: but, in looking over other long-established periodicals, I have noted their glimmers of quasi-individuality fade gradually, after about 1860, and the surrender of their attempted identities to a higher attempted organization. Some of them, expressing inter-mediateness-wide endeavour to localize the universal, or to localize self, soul, identity, entity—or positiveness or realness—held out until as far as 1880; traces findable up to 1890—and then, expressing the universal process—except that here and there in the world's history there may have been successful approximations to positiveness by "individuals"—who only then became individuals and attained to selves or souls of their own—surrendered, submitted, became parts of a higher organization's attempt to individualize or systematize into a complete thing, or to localize the universal or the attributes of the universal. After the death of Richard Proctor, whose occasional illiberalities I'd not like to emphasize too much, all succeeding volumes of *Knowledge* have yielded scarcely an unconventionality. Note the great number of times that the *American Journal of Science* and the *Report of the British Association* are quoted: note that, after, say, 1885, they're scarcely mentioned in these inspired but illicit pages—as by hypnosis and inertia, we keep on saying.

About 1880.

Throttle and disregard.

But the coercion could not be positive, and many of the excommunicated continued to creep in; or, even to this day, some of the strangled are faintly breathing.

Some of our data have been hard to find. We could tell stories of great labour and fruitless quests that would, though perhaps imperceptibly, stir the sympathy of a Mr. Symons. But, in this matter of concurrence of earthquakes with aerial phenomena, which are as unassociable with earthquakes, if internally caused, as falls of sand on convulsed small boys full of sour apples, the abundance of so-called evidence is so great that we can only sketchily go over the data, beginning with Robert Mallet's Catalogue (*Rept. Brit. Assoc.*, 1852), omitting some extraordinary instances, because they occurred before the eighteenth century:

Earthquake "preceded" by a violent tempest, England, Jan. 8, 1704—"preceded" by a brilliant meteor, Switzerland, Nov. 4, 1704—"luminous cloud, moving at high velocity, disappearing behind the horizon", Florence, Dec. 9, 1731—"thick mists in the air, through which a dim light was seen: several weeks before the shock, globes of light had been seen in the air", Swabia, May 22, 1732—rain of earth, Carpentras, France, Oct. 18, 1737—a black cloud, London, March 1, 1750—violent storm and a strange star of octagonal shape, Slavange, Norway, April 15, 1752—balls of fire from a streak in the sky, Augermannland, 1752—numerous meteorites, Lisbon, Oct. 15, 1755–"terrible tempests" over and over—"falls of hail" and "brilliant meteors" instance after instance—"an immense globe". Switzerland, Nov. 2, 1761—oblong, sulphurous cloud, Germany, April, 1767—extraordinary mass of vapour, Boulogne, April, 1780—heavens obscured by a dark mist, Grenada, Aug. 7, 1804—"strange, howling noises in the air, and large spots obscuring the sun", Palermo, Italy, April 16, 1817—"luminous meteor moving in the same direction as the shock", Naples, Nov. 22, 1821—fire ball appearing in the sky: apparent size of the moon. Thuringerwald, Nov. 28, 1831.

And, unless you be polarized by the New Dominant, which is calling for recognition of multiplicities of external things, as a Dominant, dawning new over Europe in 1492, called for recognition of terrestrial externality to Europe—unless you have this contact with the new, you have no affinity for these data—beans that drop from a magnet—irreconcilables that glide from the mind of a Thomson—

Or my own acceptance that we do not really think at all; that we correlate around super-magnets that I call Dominants—a spiritual Dominant in one age, and responsively to it up spring monasteries, and the stake and the cross are its symbols: a Materialist Dominant, and up spring laboratories, and microscopes and telescopes and crucibles are its icons—that we're nothing but iron filings relatively to a succession of magnets that displace preceding magnets.

With no soul of your own, and no soul of my own—except that some day some of us may no longer be Intermediatisms, but may hold out against the cosmos that once upon a time thousands of fishes were cast from one pail of water—we have psycho-valency for these data, if we're obedient slaves to the New Dominant, and repulsion to them, if we're mere correlates to the Old Dominant. I'm a soulless and selfless correlate to the New Dominant, myself: I see what I have to see. The only inducement I can hold out, in my attempt to rake up disciples, is that some day the New will be fashionable: the new correlates will sneer at the old correlates. After all, there is some inducement to that—and I'm not altogether sure it's desirable to end up as a fixed star.

As a correlate to the New Dominant, I am very much impressed with some of these data—the luminous object that moved in the same direction as a earthquake—it seems very acceptable that a quake followed this thing as it passed near this earth's surface. The streak that was seen in the sky—or only a streak that was visible of another world—and objects, or meteorites, that were shaken down from it. The quake at Carpentras, France: and that, above Carpentras, was a smaller world, more violently quaked, so that earth was shaken down from it.

But I like best the super-wolves that were seen to cross the sun during the earthquake at Palermo.

They howled.

Or the loves of the worlds. They call they feel for one another. Then try to move closer and howl when they get there.

The howls of the planets.

I have discovered a new unintelligibility.

In the *Edinburgh New Philosophical Journal*—have to go away back to 1841—days of less efficient strangulation—Sir David Milne lists phenomena of quakes in Great Britain. I pick out a few that indicate to me that other worlds were near this earth's surface:

Violent storm before a shock of 1703—ball of fire "preceding", 1750—a large ball of fire seen upon day following a quake, 1755—"uncommon phenomenon in the air: a large luminous body, bent like a crescent, which stretched itself over the heavens", 1816—vast ball of fire,

1750—black rains and black snows, 1755—numerous instances of upward projection—or upward attraction?—during quakes—"preceded by a cloud, very black and lowering", 1795—fall of black powder, preceding a quake, by six hours, 1837.

Some of these instances seem to me to be very striking—a smaller world: it is greatly racked by the attraction of this earth—black substance is torn down from it—not until six hours later, after an approach still closer, does the earth suffer perturbation. As to the extraordinary spectacle of a thing, world, super-construction, that was seen in the sky, in 1816, I have not yet been able to find out more. I think that here our acceptance is relatively sound: that this occurrence was tremendously of more importance than such occurrence as, say, transits of Venus, upon which hundreds of papers have been written—that not another mention have I found, though I have not looked so especially as I shall look for more data—that all but undetailed record of this occurrence was suppressed.

Altogether we have considerable agreement here between data of vast masses that do not fall to this earth, but from which substances fall, and data of fields of ice from which ice may fall, but from which water may drip. I'm beginning to modify: that, at a distance from this earth, gravitation has more effect than we have supposed, though less effect than the dogmatists suppose and "prove". I'm coming out stronger for the acceptance of a Neutral Zone—that is earth, like other magnets, has a neutral zone, in which is the Super-Sargasso Sea, and in which other worlds may be buoyed up, though projecting parts may be subject to this earth's attraction—

But my preference:

Here are the data.

I now have one of the most interesting of the new correlates. I think I should have brought it in before, but, whether out of place here, because not accompanied by earthquake, or not, we'll have it. I offer it as an instance of an eclipse, by a vast, dark body, that has been seen and reported by an astronomer. The astronomer is M. Lias: the phenomena was seen by him, at Pernambuco, April 11, 1860.

Comptes Rendus, 50–1197:

It was about noon—sky cloudless—suddenly the light of

the sun was diminished. The darkness increased, and, to illustrate its intensity, we are told that the planet Venus shone brilliant. But Venus was of low visibility at this time. The observation that burns incense to New Dominant is:

That around the sun appeared a corona.

There are many other instances that indicate proximity of other worlds during earthquakes. I note a few—quakes and an object in the sky, called "a large, luminous meteor" (*Quar. Jour. Roy. Inst.*, 5–132); luminous body in the sky, earthquake, and fall of sand, Italy, Feb. 12 and 13, 1870 (*La Science Pour Tous*, 15–159); many reports upon luminous object in the sky and earthquake, Connecticut, Feb. 27, 1883 (*Monthly Weather Review*, February, 1883); luminous object, or meteor, in the sky, fall of stones from the sky, and earthquake, Italy, Jan. 20, 1891 (*L'Astronomie*, 1891–154); earthquake and prodigious number of luminous bodies, or globes, in the air, Boulogne, France, June 7, 1779 (Sestier "*La Foudre*" 1–169); earthquake at Manila, 1863, and "curious luminous appearance in the sky" (Ponton, *Earthquakes*. p. 124).

The most probable appearance of fishes during an earthquake is that of Riobamba. Humboldt sketched one of them, and it's an uncanny-looking thing. Thousands of them appeared upon the ground during this tremendous earthquake. Humboldt says that they were cast up from subterranean sources. I think not myself, and have data for thinking not, but there'd be such a row arguing back and forth that it's simpler to consider a clearer instance of the fall of living fishes from the sky during an earthquake. I can't quite accept myself, whether a large lake, and all the fishes in it, was torn down from some other world, or a lake in the Super-Sargasso Sea, distracted between two pulling worlds, was dragged down to this earth—

Here are the data:

La Science Pour Tous, 6–191:

Feb. 16, 1861. An earthquake at Singapore. Then came an extraordinary downpour of rain—or as much water as any good-sized lake would consist of. For three days this rain or this fall of water came down in torrents. In pools on the ground, formed by this deluge, great numbers of fishes were found. The writer says that he had, himself, seen nothing but water fall from the sky. Whether I'm emphasizing what a deluge it was or not, he says that so terrific had been the downpour that he had not been able to see three steps away from him. The natives said that the fishes had fallen from the sky. Three days later the pools dried up and many dead fishes were found, but, in the first place—though that's an expression for which we have an instinctive dislike—the fishes had been active and uninjured. Then follows material for another of our little studies in the phenomena of disregard. A psychotropism here is mechanically to take pen in hand and mechanically write that fishes found on the ground after a heavy rainfall came from overflowing streams. The writer of the account says that some of the fishes had been found in his courtyard, which was surrounded by high walls—paying no attention to this, correspondent (*La Science Pour Tous*, 6–317) explains that in the heavy rain a body of water had probably overflowed, carrying fishes with it. We are told by the first writer that these fishes of Singapore were of a species that was very abundant near Singapore. So I think, myself, that a whole lakeful of them had been shaken down from the Super-Sargasso Sea, under the circumstances we have thought of. However, if appearance of strange fishes after an earthquake be more pleasing in the sight, or to the nostrils, of the New Dominant, we faithfully and piously supply that incense—An account of the occurrence at Singapore was read by M. de Castelnau, before the French Academy. M. de Castelnau recalled that, upon a former occasion, he had submitted to the Academy the circumstance that fishes of a new species had appeared at the Cape of Good Hope after an earthquake.

It seems proper, and it will give lustre to the new orthodoxy, now to have an instance in which, not merely quake and fall of rocks, or meteorites, or quake and either eclipse of luminous appearances in the sky have occurred, but in which are combined all the phenomena, one or more of which, when accompanying earthquake, indicate, in our acceptance, the proximity of another world. This time a longer duration is indicated than in other instances.

It seems proper, and it will give lustre to the new orthodoxy, in the *Canadian Institute Proceedings*, 2–7–198, there is an account, by the Deputy Commissioner at Dhurmsalla, of the extraordinary Dhurmsalla meteorite—coated with ice. But the combination of events related by him is still more extraordinary:

That within a few months of the fall of this meteorite there had been a fall of live fishes at Benares, a shower of red substance at Furruckabad, a dark spot observed on the disc of the sun, an earthquake, "an unnatural darkness of some duration", and a luminous appearance in the sky that looked like an aurora borealis—

But there's more to this climax:

We are introduced to a new order of phenomena:

Visitors.

The Deputy Commissioner writes that, in the evening, after the fall of the Dhurmsalla meteorite, or mass of stone covered with ice, he saw lights. Some of them were not very high. They appeared and went out and reappeared. I have read many accounts of the Dhurmsalla meteorite—July 28, 1860—but never in any other of them a mention of this new correlate—something as out of place in the nineteenth century as would have been an aeroplane—the invention of which would not, in our acceptance, have been permitted, in the nineteenth century, though adumbrations to it were permitted. This writer says that the light moved like fire balloons, but:

"I am sure that they were neither fire balloons, lanterns, nor bonfires, or any other thing of that sort, but bona fide lights in the heavens."

It's a subject for which we shall have to have a separate expression—trespassers upon territory to which something else has a legal right—perhaps someone lost a rock, and he and his friends came down looking for it, in the evening—or secret agents, or emissaries, who had an appointment with certain esoteric ones near Dhurmsalla—things or beings coming down to explore, and unable to stay down long—

In a way, another strange occurrence during an earthquake is suggested. The ancient Chinese tradition—the marks like hoof marks in the ground. We have thought—with a low degree of acceptance—of another world that may be in secret communication with certain esoteric ones of this earth's inhabitants—and of messages in symbols like hoof marks that are sent to some receptor, or special hill, upon this earth—and of messages that at times miscarry.

This other world comes close to this world—there are quakes—but advantage of proximity is taken to send a message—the message, designed for a receptor in India, perhaps, or in Central Europe, miscarries all the way to England—marks like the marks of the Chinese tradition are found upon a beach, in Cornwall, after an earthquake—

Phil. Trans., 50–500:

After the quake of July 15, 1757, upon the sands of Penzance, Cornwall, in an area of more than 100 square yards, were found marks like hoof prints, except that they were not crescentic. We feel a similarity, but note an arbitrary disregard of our own, this time. It seems to us that marks described as "little cones surrounded by basins of equal diameter" would be like hoof prints, if hoof prints complete circles. Other disregards are that there were black specks on the tops of cones, as if something, perhaps gaseous, had issued from them; that from one of these formations came a gush of water as thick as a man's wrist. Of course the opening of springs is common in earthquakes—but we suspect, myself, that the Negative Absolute is compelling us to put in this datum and its disorders.

There's another matter in which the Negative Absolute seems to work against us. Though to super-chemistry, we have introduced the principle of celestio-metathesis, we have no good data of exchange of substances during proximities. The data are all of falls and not of upward translations. Of course upward impulses are common during earthquakes, but I haven't a datum upon a tree or a fish or a brick or a man that ever did go up and stay up and that never did come down again. Our classic of the horse and barn occurred in what was called a whirlwind.

It is said that, in an earthquake in Calabria, paving stones shot up far in the air.

The writer doesn't specifically say that they came down again, but something seems to tell me they did.

The corpses of Riobamba.

Humboldt reported that, in the quake of Riobamba, "bodies were torn upward from graves"; that "the vertical motion was so strong that bodies were tossed several hundred feet in the air".

I explain.

I explain that, if in the centre of greatest violence of an earthquake, anything ever has gone up, and has kept on going up, the thoughts of the nearest observers were very likely upon other subjects.

The quay of Lisbon.

We are told that it went down.

A vast throng of persons ran to the quay for refuge. The city of Lisbon was in profound darkness. The quay and all the people on it disappeared. If it and they went down—not a single corpse, not a shred of clothing, not a plank of the quay, nor so much as a splinter of it ever floated to the surface.

Chapter 18

The New Dominant.

I mean "primarily" all that opposes Exclusionism—

That Development or Progress or Evolution is Attempt to Positivize, and is a mechanism by which a positive existence is recruited—that what we call existence is a womb of infinitude, and is itself only incubatory—that eventually all attempts are broken down by the falsely excluded. Subjectively, the breaking down, is aided by our own sense and narrow limitations. So the classic and academic artists wrought positivist paintings, and expressed the only ideal that I am conscious of, though we so often hear of "ideals" instead of different manifestations, artistically, scientifically, theologically, politically, of the One Ideal. They sought to satisfy, in its artistic aspect, cosmic craving for unity or completeness, sometimes called harmony, called beauty in some aspects. By disregard they sought completeness. But the light-effects that they disregarded, and their narrow confinement to standardized subjects brought on the revolt of the Impressionists. So the Puritans tried to systematize, and they disregarded physical needs, or vices, or relaxations: they were invaded and overthrown when their narrowness became obvious and intolerable. All things strive for positiveness, for themselves, or for quasi-systems of which they are parts. Formality and the mathematic, the regular and the uniform are aspects of the positive state—but the Positive is the Universal—so all attempted positiveness that seems to satisfy in the aspects of formality and regularity, sooner or later disqualifies in the aspect of wideness or universalness. So there is revolt against the science of today, because the formulated utterances that were regarded as final truths in a past generation, are now seen to be insufficiencies. Every pronouncement that has opposed our own acceptances has been found to be a composition like any academic painting: something that is arbitrarily cut off from relations with environment, or framed off from interfering and disturbing data, or outlined with disregards. Our own attempt has been to take in the

258

included, but also to take in the excluded into wider expressions. We accept, however, that for every one of our expressions there are irreconcilables somewhere—that final utterance would include all things. However, of such is the gossip of angels. The final is unutterable in quasi-existence, where to think is to include but also to exclude, or be not final. If we admit that for every opinion we have expressed, there must somewhere be an irreconcilable, we are Intermediatists and not positivists; not even higher positivists. Of course it may be that some day we shall systematize and dogmatize and refuse to think of anything that we may be accused of disregarding, and believe instead of merely accepting: then, if we could have a wider system, which would acknowledge no irreconcilables we'd be higher positivists. So long as we only accept, we are not higher positivists, but our feeling is that the New Dominant, even though we have thought of it only as another enslavement, will be the nucleus for higher positivism—and that it will be the means of elevating into infinitude a new batch of fixed stars—until, as a recruiting instrument, it, too, will play out, and will give way to some new medium for generating absoluteness. It is our acceptance that all astronomers of today have lost their souls, or, rather, all chance of attaining Entity, but that Copernicus and Kepler and Galileo and Newton, and, conceivably, Leverrier are now fixed stars. Some day I shall attempt to identify them. In all this, I think we're quite a Moses. We point out the Promised Land, but, unless we be cured of our Intermediatism, will never be reported in *Monthly Notices*, ourself.

In our acceptance, Dominants, in their succession, displace preceding Dominants not only because they are nearly positive, but because the old Dominants, as recruiting mediums, play out. Our expression is that the New Dominant, of Wider Inclusions, is now manifesting throughout the world, and that the old Exclusionism is everywhere breaking down. In physics Exclusionism is breaking down by its own researches in radium, for instance, and in its speculations upon electrons, or its merging away into metaphysics, and by the desertion that has been going on for many years, by such men as Gurney, Crookes, Wallace, Flammarion, Lodge, to formerly disregarded phenomena—no longer called "spiritualism" but now "psychic research". Biology is in chaos: conventional

Darwinites mixed up with mutationists and orthogenesists and followers of Wisemann, who take from Darwinism one of its pseudo-bases, and nevertheless try to reconcile their heresies with orthodoxy. The painters are metaphysicians and psychologists. The breaking down of Exclusionism in China and Japan and in the United States has astonished History. The science of astronomy is going downward so that, though Pickering, for instance, did speculate upon a Trans-Neptunian planet, and Lowell did try to have accepted heretical ideas as to marks on Mars, attention is now minutely focused upon such technicalities as variations in shades of Jupiter's fourth satellite. I think that, in general acceptance, over-refinement indicates decadence.

I think that the stronghold of Inclusionism is in aeronautics. I think that the stronghold of the Old Dominant, when it was new, was in the invention of the telescope. Or that coincidentally with the breakdown of Exclusionism appears the means of finding out—whether there are vast aerial fields of ice and floating lakes full of frogs and fishes or not—where carved stones and black substances and great quantities of vegetable matter and flesh, which may be dragons' flesh, come from—whether there are inter-planetary trade routes and vast areas devastated by Super-Tamerlanes—whether sometimes there are visitors to this earth—who might be pursued and captured and questioned.

Chapter 19

I have industriously sought data for an expression upon birds, but the prospecting has not been very quasi-satisfactory. I think I rather emphasize our industriousness, because a charge likely to be brought against the attitude of Acceptance is that one who accepts must be one of languid interest and little application of energy. It doesn't seem to work out: we are very industrious. I suggest to some of our disciples that they look into the matter of messages upon pigeons, of course attributed to earthly owners, but said to be undecipherable. I'd do it, ourselves, only that would be selfish. That's more of the Intermediatism that will keep us out of the firmament: Positivism is absolute egoism. But look back in the time of André's Polar Expedition. Pigeons that would have no publicity ordinarily, were often reported at that time.

In the *Zoologist*, 3–18–21, is recorded an instance of a bird (puffin) that had fallen to the ground with a fractured head. Interesting, but mere speculation—but what solid object, high in the air, had that bird struck against?

Tremendous red rain in France, Oct. 16 and 17, 1846; great storm at the time, and red rain supposed to have been coloured by matter swept up from this earth's surface, and then precipitated (*Comptes Rendus*, 23–832). But in *Comptes Rendus*, 24–625, the description of this red rain differs from one's impression of red, sandy or muddy water. It is said that this rain was so vividly red and so blood-like that many persons in France were terrified. Two analyses are given (*Comptes Rendus*, 24–812). One chemist notes a great quantity of corpuscles—whether blood-like corpuscles or not—in the matter. The other chemist sets down organic matter at 35 per cent. It may be that an inter-planetary dragon had been slain somewhere, or that this red fluid, in which were many corpuscles, came from something not altogether pleasant to contemplate, about the size of the Catskill Mountains, perhaps—but the present datum is that with his substance, larks, quail, ducks, and water hens, some of them alive, fell at Lyons and Grenoble and other places.

I have notes upon other birds that have fallen from the sky, but unaccompanied by the red rain that makes the fall of birds in France peculiar, and very peculiar, if it be accepted that the red substance was extra-mundane. The other notes are upon birds that have fallen from the sky, in the midst of storms, or of exhausted, but living, birds, falling not far from a storm-area. But now we shall have an instance for which I can find no parallel: fall of dead birds, from a clear sky, far-distant from any storm to which they could be attributed—so remote from any discoverable storm that—

My own notion is that, in the summer of 1896, something, or some beings, came as near to this earth as they could, upon a hunting expedition; that, in the summer of 1896, an expedition of super-scientists passed over the earth, and let down a dragnet—and what would it catch, sweeping through the air, supposing it to have reached not quite to this earth?

In the *Monthly Weather Review*, May, 1917, W. L. McAtee quotes from the Baton Rouge correspondence to the *Philadelphia Times:*

That, in the summer of 1896, into the streets of Baton Rouge, La., and from a "clear sky", fell hundreds of dead birds. There were wild ducks and cat birds, woodpeckers, and "many birds of strange plumage", some of them resembling canaries.

Usually one does not have to look very far from any place to learn of a storm. But the best that could be done in this instance was to say:

"There had been a storm on the coast of Florida."

And, unless he have psycho-chemic repulsion for the explanation, the reader feels only momentary astonishment that dead birds from a storm in Florida should fall from an unstormy sky in Louisiana, and with his intellect greased like the plumage of a wild duck, the datum then drops off.

Our greasy, shiny brains. That they may be of some use after all; that other modes of existence place a high value upon them as lubricants; that we're hunted for them; a hunting expedition to this earth—the newpapers report a tornado.

If from a clear sky, or a sky in which there were no driven

clouds, or other evidences of still-continuing wind-power—or, if from a storm in Florida, it could be accepted that hundreds of birds had fallen far away, in Louisiana, I conceive, conventionally, of heavier objects having fallen in Alabama, say, and of the fall of still heavier objects still nearer the origin in Florida.

The sources of information of the Weather Bureau are widespread.

It has no records of such falls.

So a dragnet that was let down from above somewhere—

Or something that I learned from the more scientific of the investigators of psychic phenomena:

The reader begins their works with prejudice against telepathy and everything else of psychic phenomena. The writers deny spirit-communication, and say that the seeming data are data of "only telepathy". Astonishing instances of seeming clairvoyance—"only telepathy". After a while the reader finds himself agreeing that it's only telepathy—which, at first, had been intolerable to him.

So maybe, in 1896, a super-dragnet did not sweep through this earth's atmosphere, gathering up all the birds within its field, the meshes then suddenly breaking—

Or that the birds of Baton Rouge were only from the Super-Sargasso Sea—

Upon which we shall have another expression. We thought we'd settled that, and we thought we'd established that, but nothing's ever settled, and nothing's ever established, in a real sense, if, in a real sense, there is nothing in quasiness.

I suppose there had been a storm somewhere, the storm in Florida, perhaps, and many birds had been swept upward into the Super-Sargasso Sea. It has frigid regions and it has tropical regions—that birds of diverse species had been swept upward, into an icy region, where, huddling together for warmth, they had died. Then later, they had been dislodged—meteor coming along—boat—bicycle—dragon—don't know what did come along—something dislodged them.

So leaves of trees, carried up there in whirlwinds, staying there years, ages, perhaps only a few months, but then falling to this earth at an unseasonable time for dead leaves—fishes carried up there, some of them dying and drying, some of them living in volumes of water that are in abundance up there, or that fall sometimes in the deluges that we call "cloudbursts".

The astronomers won't think kindly of us, and we haven't done anything to endear ourselves to the meteorologists —but we're weak and mawkish Intermediatists— several times we've tried to get the aeronauts with us— extraordinary things up there: things that curators of museums would give up all hope of ever being fixed stars, to obtain: things left over from whirlwinds of the time of the Pharoahs, perhaps: or that Elijah did go up in the sky in something like a chariot, and may not be Vega, after all, and that there may be a wheel or so left of whatever he went up in. We basely suggest that it would bring a high price—but sell soon, because after a while there'd be thousands of them hawked around—

We weakly drop a hint to the aeronauts.

In the *Scientific American*, 33–197, there is an account of some hay that fell from the sky. From the circumstances we incline to accept that this hay went up, in a whirlwind, from this earth, in the first place, reached the Super-Sargasso Sea, and remained there a long time before falling. An interesting point in this expression is that usual attribution to a local and coinciding whirlwind, and identification of it—and then data that make the local whirlwind unacceptable—

That, upon July 27, 1875, small masses of damp hay had fallen at Monkstown, Ireland. In the *Dublin Daily Express*, Dr. J. W. Moore had explained: he had found a nearby whirlwind, to the south of Monkstown, that coincided. But, according to the *Scientific American*, a similar fall had occurred near Wrexham, England, two days before.

In November, 1918, I made some studies upon light objects thrown into the air. Armistice-day. I suppose I should have been more emotionally occupied, but I made notes upon torn-up papers thrown high in the air from windows of office buildings. Scraps of paper did stay together for a while. Several minutes, sometimes.

Cosmos, 3–4–574:

That, upon the 10th of April, 1869, at Autriche (Indre-et-Loire) a great number of oak leaves—enormous segregation of them—fell from the sky. Very calm day. So little wind that the leaves fell almost vertically. Fall lasted about ten minutes.

Flammarion, in *The Atmosphere*, p. 412, tells this story. He has to find a storm.

He does find a squall—but it had occurred upon April 3rd.

Flammarion's two incredibilities are—that leaves could remain a week in the air: that they could stay together a week in the air.

Think of some of your own observations upon papers thrown from an aeroplane.

Our one incredibility:

That these leaves had been whirled up six months before, when they were common on the ground, and had been sustained, of course not in the air, but in a region gravitationally inert; and had been precipitated by disturbances of April rains.

I have no records of leaves that have so fallen from the sky in October or November, the season when one might expect dead leaves to be raised from one place and precipitated somewhere else. I emphasize that this occurred in April.

La Nature, 1889-2-94:

That, upon April 19, 1889, dried leaves, of different species, oak, elm, etc., fell from the sky. This day too, was a calm day. The fall was tremendous. The leaves were seen to fall fifteen minutes but, judging from the quantity on the ground, it is the writer's opinion that they had already been falling half an hour. I think that the geyser of corpses that sprang from Riobamba toward the sky must have been an interesting sight. If I were a painter, I'd like that subject. But this cataract of dried leaves, too, is a study in the rhythms of the dead. In this datum, the point most agreeable to us is the very point that the writer in *La Nature* emphasizes. Windlessness. He says that the surface of the Loire was "absolutely smooth". The river was strewn with leaves as far as he could see.

L'Astronomie, 1894-194:

That, upon the 7th of April, 1894, dried leaves fell at Clairvaux and Outre-Aube, France. The fall is described as prodigious. Half an hour. Then, upon the 11th, a fall of dried leaves occurred at Pontcarré.

It is in this recurrence that we found some of our opposition to the conventional explanation. The Editor

(Flammarion) explains. He says that the leaves had been caught up in a cyclone which had expended its force; that the heavier leaves had fallen first. We think that that was all right for 1894, and that it was quite good enough for 1894. But, in these more exacting days, we want to know how wind-power insufficient to hold some leaves in the air could sustain others four days.

The factors of this expression are unseasonableness, not for dried leaves, but for prodigious numbers of dried leaves; direct fall, windlessness, month of April, and localization in France. The factor of localization is interesting. Not a note have I upon fall of leaves from the sky, except these notes. Were the conventional explanation, or "old correlate" acceptable, it would seem that similar occurrences in other regions should be as frequent as in France. The indication is that there may be quasi-permanent undulations in the Super-Sargasso Sea, or a pronounced inclination toward France—

Inspiration:

That there may be a nearby complementary to this world, where autumn occurs at the time that is springtime here.

Let some disciple have that.

But there may be a dip toward France, so that leaves that are born high there, are more likely to be held in suspension than high-flying leaves elsewhere. Some other time I shall take up Super-geography, and be guilty of charts, I think, now, that the Super-Sargasso Sea is an oblique belt, with changing ramifications, over Great Britain, France, Italy, and on to India. Relatively to the United States I am not very clear but think especially of the Southern States.

The preponderance of our data indicates frigid regions aloft. Nevertheless such phenomena as putrefaction have occurred often enough to make super-tropical regions, also, acceptable. We shall have one more datum upon the Super-Sargasso Sea. It seems to me that, by this time, our requirements of support and reinforcement and agreement have been quite as rigorous for acceptance as ever for belief: at least for full acceptance. By virtue of mere acceptance, we may, in some later book, deny the Super-Sargasso Sea, and find that our data to some other complementary world instead—or the moon—and have abundant data for accepting that the moon is not more than twenty or thirty miles away. However, the Super-Sargasso Sea functions very well as a nucleus around which to gather data that oppose Exclusionism. That is our main motive: to oppose Exclusionism.

Or our agreement with cosmic processes. The climax of our general expression upon the Super-Sargasso Sea. Coincidentally appears something that may overthrow it later.

Notes and Queries, 8–12–228:

That in the province of Macerata, Italy (summer of 1897?) an immense number of small, blood-coloured clouds covered the sky. About an hour later a storm broke, and myriad seeds fell to the ground. It is said that they were identified as products of a tree found only in Central Africa and the Antilles.

If—in terms of conventional reasoning—these seeds had been high in the air, they had been in a cold region. But it is our acceptance that these seeds had, for a considerable time, been in a warm region, and for a time longer than is attributable to suspension by wind-power:

"It is said that a great number of the seeds were in the first stage of germination."

Chapter 20

The New Dominant.

Inclusionism.

In it we have a pseudo-standard.

We have a datum, and we give it an interpretation, in accordance with our pseudo-standard. At present we have not the delusions of Absolutism that may have translated some of the positivists of the nineteenth century to heaven. We are Intermediatists—but feel a lurking suspicion that we may some day solidify and dogmatize and illiberalize into higher positivists. At present we do not ask whether something be reasonable or preposterous, because we recognize that by reasonableness and preposterousness are meant agreement and disagreement with a standard—which must be a delusion—though not absolutely, of course—and must some day be displaced by a more advanced quasi-delusion. Scientists in the past have taken the positivist attitude—is this or that reasonable or unreasonable? Analyse them and we find that they meant relatively to a standard, such as Newtonism, Daltonism, Darwinism, or Lyellism. But they have written and spoken and thought as if they could mean real reasonableness and real unreasonableness.

So our pseudo-standard is Inclusionism, and, if a datum be a correlate to a more widely inclusive outlook as to this earth and its externality and relations with externality, its harmony with Inclusionism admits it. Such was the process, and such was the requirement for admission in the days of the Old Dominant: our difference is in underlying Intermediatism, or consciousness that though we're more nearly real, we and our standards are only quasi—

Or that all things—in our intermediate state—are phantoms in a super-mind in a dreaming state—but striving to awaken to realness.

Though in some respects our own Intermediatism is unsatisfactory, our underlying feeling is—

That in a dreaming mind awakening is accelerated—if phantoms in that mind know that they're only phantoms in a dream. Of course, they too are quasi, or—but in a relative sense—they have an essence of what is called realness. They are derived from experience or from sense-relations, even though grotesque distortions. It seems acceptable that a table that is seen when one is awake is more nearly real than a dreamed table, which, with fifteen or twenty legs, chases one.

So now, in the twentieth century, with a change of terms, and a change in underlying consciousness, our attitude toward the New Dominant is the attitude of the scientists of the nineteenth century to the Old Dominant. We do not insist that our data and interpretations shall be as shocking, grotesque, evil, ridiculous, childish, insincere, laughable, ignorant to nineteenth-centuryites, as were their data and interpretations correlate. If they do they are acceptable, perhaps only for a short time, or as nuclei, or scaffolding, or preliminary sketches, or as gropings and tentativenesses. Later of course, when we cool off and harden and radiate into space most of our present mobility, which expresses in modesty and plasticity, we shall acknowledge no scaffoldings, gropings or tentativenesses, but think we utter absolute facts. A point in Intermediatism here is opposed to most current speculations upon Development. Usually one thinks of the spiritual as higher than the material, but, in our acceptance, quasi-existence is a means by which the absolutely immaterial materializes absolutely, and, being intermediate, is a state in which nothing is finally either immaterial or material, all objects, substances, thoughts, occupying some grade of approximation one way or the other. Final solidification of the ethereal is, to us, the goal of cosmic ambition. Positivism is Puritanism. Heat is Evil, Final Good is Absolute Frigidity. An Arctic winter is very beautiful, but I think that an interest in monkeys chattering in palm trees accounts for our own Intermediatism.

Visitors.

Our confusion here, out of which we are attempting to make quasi-order, is as great as it has been throughout this book, because we have not the positivist's delusion of homogeneity. A positivist would gather all data that seem to relate to one kind of visitors and coldly disregard all other data. I think of as many different kinds of visitors to this earth as there are visitors to New York, to a jail, to a

269

church—some persons go to church to pick pockets, for instance.

My own acceptance is that either a world or a vast super-construction—or a world, if red substances and fishes fell from it—hovered over India in the summer of 1860. Something then fell from somewhere, July 17, 1860, at Dhurmsalla. Whatever "it" was, "it" is so persistently alluded to as "a meteorite" that I look back and see that I adopted this convention myself. But in the London *Times*, Dec. 26, 1860, Syed Abdoolah, Professor of Hindustani, University College, London, writes that he had sent to a friend in Dhurmsalla, for an account of the stones that had fallen at that place. The answer:

". . . divers forms and sizes many of which bore great resemblance to ordinary cannon balls just discharged from engines of war."

It's an addition to our data of spherical objects that have arrived upon this earth. Note that they are spherical stone objects.

And, in the evening of this same day that something—took a shot at Dhurmsalla—or sent objects upon which there may be decipherable markings—lights were seen in the air—

I think, myself, of a number of things, beings, whatever they were, trying to get down, but resisted, like balloonists, at a certain altitude trying to get farther up but resisted.

Not in the least except to good positivists, or the homogeneous-minded, does this speculation interfere with the concept of some other world that is in successful communication with certain esoteric ones upon this earth, by a code of symbols that print in rock, like symbols of tele-photographers in selenium.

I think that sometimes, in favourable circumstances, emissaries have come to this earth—secret meetings—

Of course it sounds—

But:

Secret meetings—emissaries—esoteric ones in Europe, before the war broke out—

And those who suggest that such phenomena could be.

However, as to most of our data, I think of super-things that have passed close to this earth with no more interest in this earth than have passengers upon a steamship in the bottom of the sea—or passengers may have a keen interest, but circumstances of schedules and commercial requirements forbid investigation of the bottom of the sea.

Then, on the other hand, we may have data of super-scientific attempts to investigate phenomena of this earth from above—perhaps by beings from so far away that they had never been heard that something, somewhere, asserts a legal right to this earth.

Altogether, we're good intermediatists, but we can't be very good hypnotists.

Still another source of the merging away of our data:

That, upon general principles of Continuity, if super-vessels, or super-vehicles, have traversed this earth's atmosphere, there must be mergers between them and terrestrial phenomena: observations upon them must merge away into observations upon clouds and balloons and meteors. We shall begin with data that we cannot distinguish ourselves and work our way out of mergers into extremes.

In the *Observatory*, 35–168, it is said that, according to a newspaper, March 6, 1912, residents of Warmley, England, were greatly excited by something that was supposed to be "a splendidly illuminated aeroplane passing over the village". The machine was apparently travelling at a tremendous rate, and came from the direction of Bath, and went on toward Gloucester. The Editor says that it was a large, triple-headed fireball. "Tremendous indeed" he says. "But we are prepared for anything nowadays."

That is satisfactory. We'd not like to creep up stealthily, and then jump out of a corner with our data. This Editor, at least, is prepared to read—

Nature, Oct. 27, 1898:

A correspondent writes that, in the County Wicklow, Ireland, at about 6 o'clock in the evening, he had seen, in the sky, an object that looked like the moon in its three-quarter aspect. We note the shape which approximates to triangularity, and we note that in colour it is said to have been golden yellow. It moved slowly, and in about five minutes disappeared behind a mountain.

The Editor gives his opinion that the object may have been an escaped balloon.

In *Nature*, Aug. 11, 1898, there is a story, taken from the July number of the *Canadian Weather Review*, by the meteorologist, F. F. Payne: that he had seen, in the

271

Canadian sky, a large, pear-shaped object, sailing rapidly. At first he supposed that the object was a balloon, "its outline being sharply defined". But, as no cage was seen, it was concluded that it must be a mass of cloud. In about six minutes this object became less definite—whether because of increasing distance or not—"the mass became less dense, and finally it disappeared". As to cyclonic formation—"no whirling motion could be seen".

Nature, 58–294;

That, upon July 8, 1898, a correspondent had seen, at Kiel, an object in the sky, coloured red, by the sun which had set. It was about as broad as a rainbow and about twelve degrees high. "It remained in its original brightness about five minutes, and then faded rapidly, and then remained almost stationary again, finally disappearing about eight minutes after I first saw it."

In an intermediate existence, we quasi-persons have nothing to judge by because everything is its own opposite. If a hundred dollars a week be a standard of luxurious living to some persons, it is poverty to others. We have instances of three objects that were seen in the sky in a space of three months, and this concurrence seems to me to be something to judge by. Science has been built upon concurrence: so have been most of the fallacies and fanaticisms. I feel the positivism of a Leverrier, or instinctively take to the notion that all three of these observations relate to the same object. However, I don't formulate them and predict the next transit. Here's another chance for me to become a fixed star—but as usual—oh, well—

A point in Intermediatism:

That the Intermediatist is likely to be a flaccid compromiser.

Our own attitude:

Ours is a partly positive and partly negative state, or a state in which nothing is finally positive or finally negative—

But if positivism attracts you, go ahead and try: you will be in harmony with cosmic endeavour—but Continuity will resist you. Only to have appearance in quasiness is to be proportionately positive, but beyond a degree of attempted positivism, Continuity will rise to pull you back. Success, as it is called—though there is only success-failure in Intermediateness—will, in Intermediateness, be yours proportionately as you are in adjustment with its own state, or some positivism mixed with compromise and retreat. To be very positive is to be a Napoleon Bonaparte, against whom the rest of civilization will sooner or later combine. For interesting data, newspaper accounts of fate of one Dowie, of Chicago.

Intermediatism, then, is recognition that our state is only a quasi-state: it is no bar to one who desires to be positive: it is recognition that he cannot be positive and remain in a state that is positive-negative. Or that a great positivist—isolated—with no system to support him—will be crucified, or will starve to death, or will be put in jail and beaten to death—that these are the birth-pangs of translation to the Positive Absolute.

So, though positive-negative, myself, I feel the attraction of the positive pole of our intermediate state, and attempt to correlate these three data: to see them homogeneously; to think that they relate to one object.

In the aeronautic journals and in the London *Times* there is no mention of escaped balloons, in the summer or fall of 1898. In the *New York Times* there is no mention of ballooning in Canada or the United States, in the summer of 1898.

London *Times*, Sept. 29, 1885:

A clipping from the *Royal Gazette*, of Bermuda, of Sept. 8, 1885, sent to *The Times* by General Lefroy:

That, upon Aug. 27, 1885, at about 8.30 A.M., there was observed by Mrs. Adelina D. Bassett, "a strange object in the clouds, coming from the north". She called the attention of Mrs. L. Lowell to it, and they were both somewhat alarmed. However, they continued to watch the object steadily for some time. It drew nearer. It was of triangular shape, and seemed to be about the size of a pilot-boat mainsail, with chains attached to the bottom of it. While crossing the land it had appeared to descend, but, as it went out to sea, it ascended, and continued to ascend, until it was lost to sight in the clouds.

Or with such power to ascend, I don't think much myself of the notion that it was an escaped balloon, partly deflated. Nevertheless, General Lefroy, correlating with Exclusionism, attempts to give a terrestrial interpretation to this occurrence. He argues that the thing may have been a balloon that had escaped fron France or England—or the

273

only aerial thing of terrestrial origin that, even to this date of about thirty-five years later, has been thought to have crossed the Atlantic Ocean. He accounts for the triangular form by deflation—"a shapeless bag, barely able to float". My own acceptance is that great deflation does not accord with observations upon its power to ascend.

In *The Times*, Oct. 1, 1885, Charles Harding, of the R.M.S., argues that if it had been a balloon from Europe, surely it would have been seen and reported by many vessels. Whether he was as good a Briton as the General or not, he shows awareness of the United States—or that the thing may have been a partly collapsed balloon that had escaped from the United States.

General Lefroy wrote to *Nature* about it (*Nature*, 33–99), saying—whatever his sensitivenesses, may have been—that the columns of *The Times* were "hardly suitable" for such a discussion. If, in the past, there had been more persons like General Lefroy, we'd have better than the mere fragments of data that in most cases are too broken up very well to piece together. He took the trouble to write to a friend of his, W. H. Gosling, of Bermuda—who also was an extraordinary person. He went to the trouble of interviewing Mrs. Bassett and Mrs. Lowell. Their description to him was somewhat different:

An object from which nets were suspended—

Deflated balloon, with its network hanging from it—

A super-dragnet?

That something was trawling overhead?

The birds of Baton Rouge.

Mr. Gosling wrote that the item of chains, or suggestion of a basket that had been attached, had originated with Mr. Bassett, who had not seen the object. Mr. Gosling mentioned a balloon that had escaped from Paris in July. He tells of a balloon that fell in Chicago, September 17, or three weeks later than the Bermuda object.

It's one incredibility against another, with disregards and convictions governed by whichever of the two Dominants looms stronger in each reader's mind. That he can't think for himself any more than I can is understood.

My own correlates:

I think that we're fished for. It may be that we're highly esteemed by super-epicures somewhere. It makes me more cheerful when I think that we may be of some use after all. I think that dragnets have often come down and have been mistaken for whirlwinds and waterspouts. Some accounts of seeming structure in whirlwinds and waterspouts are astonishing. And I have data that, in this book, I can't take up at all—mysterious disappearances. I think we're fished for. But this is a little expression on the side: relates to trespassers; has nothing to do with the subject that I shall take up at some other time—or our use to some other mode of seeming that has a legal right to us.

Nature, 33–137:

"Our Paris correspondent writes that in relation to the balloon which is said to have been seen over Bermuda, in September, no ascent took place in France which can account for it."

Last of August: not September. In the London *Times* there is no mention of balloon ascents in Great Britain, in the summer of 1885, but mention of two ascents in France. Both balloons had escaped. In *Aéronaute*, August, 1885, it is said that these balloons had been sent up from fêtes of the fourteenth of July—44 days before the observation at Bermuda. The aeronauts were Gower and Eloy. Gower's balloon was found floating on the ocean, but Eloy's balloon was not found. Upon the 17th of July it was reported by a sea captain: still in the air; still inflated.

But this balloon of Eloy's was a small exhibition balloon, made for short ascents from fêtes and fair grounds. In *La Nature*, 1885–2–131, it is said that it was a very small balloon, incapable of remaining long in the air.

As to contemporaneous ballooning in the United States, I find only one account: an ascent in Connecticut, July 29, 1885. Upon leaving this balloon, the aeronauts had pulled the "rip cord", "turning it inside out". (*New York Times*, Aug. 10, 1885.)

To the Intermediatist, the accusation of "anthropomorphism" is meaningless. There is nothing in anything that is unique or positively different. We'd be materialists were it not quite as rational to express the material terms of the immaterial as to express the immaterial in terms of the material. Oneness of allness in quasiness. I will engage to write the formula of any novel in psycho-chemic terms, or draw its graph in psycho-mechanic terms: or write, in romantic terms, the circumstances and sequences of any chemic or electric or magnetic reaction: or express any historic event in algebraic terms—or see Boole and Jeavons for economic situations expressed algebraically.

I think of the Dominants as I think of persons—not meaning that they are real persons—not meaning that we are real persons—

Or the Old Dominant and its jealousy, and its suppression of all things and thoughts that endangered its supremacy. In reading discussions of papers, by scientific societies, I have often noted how, when they approached forbidden—or irreconcilable—subjects, the discussions have often been led astray—as if purposefully—as if by something directive, hovering over them. Of course I mean only the Spirit of all Development. Just so, in any embryo, cells that would tend to vary from appearances of their era are compelled to correlate.

In *Nature*, 90–169, Charles Tilden Smith writes that, at Chisbury, Wiltshire, England, April 8, 1912, he saw something in the sky—

"—unlike anything that I had ever seen before".

"Although I have studied the skies for many years I have never seen anything like it."

He saw two stationary dark patches upon clouds.

The extraordinary part:

They were stationary upon clouds that were rapidly moving.

They were fan-shaped—or triangular—and varied in size, but kept the same position upon different clouds as cloud after cloud came along. For more than half an hour Mr. Smith watched these dark patches—

His impression as to the one that appeared first:

That it was "really a heavy shadow cast upon a thin veil of clouds by some unseen object away to the west, which was intercepting the sun's rays".

Upon page 244 of this volume of *Nature*, is a letter from another correspondent, to the effect that similar shadows are cast by mountains upon clouds, and that no doubt Mr. Smith was right in attributing the appearance to "some unseen object, which was intercepting the sun's rays". But the Old Dominant that was a jealous Dominant, and the wrath of the Old Dominant against such an irreconcilability as large, opaque objects in the sky, casting down shadows upon clouds. Still the Dominants are suave very often, or are not absolute gods, and the way attention was led away from this subject is an interesting study in quasi-divine bamboozlement. Upon page 268, Charles P. Cave, the meteorologist, writes that, upon April 5 and 8, at Ditcham Park, Petersfield, he had observed a similar appearance, while watching some pilot balloons—but he describes something not in the least like a shadow on clouds, but a stationary cloud—the inference seems to be that the shadows at Chisbury may have been shadows of pilot balloons. Upon page 322, another correspondent writes upon shadows cast by mountains; upon page 348 someone else carries on the divergence by discussing this third letter: then someone takes up the third letter mathematically; and then there is a correction of error in this mathematical demonstration—I think it looks very much like what I think it looks like.

But the mystery here:

That the dark patches at Chisbury could not have been cast by stationary pilot balloons that were to the west, or that were between clouds and the setting sun. If, to the west of Chisbury, a stationary object were high in the air, intercepting the sun's rays, the shadow of the stationary object would not have been stationary, but would have moved higher with the setting of the sun.

I have to think of something that is in accord with no other data whatsoever:

A luminous body—not the sun—in the sky—but, because of some unknown principle or atmospheric condition, its light extended down only to the clouds; that from it were suspended two triangular objects, like the object that was seen in Bermuda; that it was this light that fell short of the earth that these objects intercepted; that the objects were drawn up and lowered from something overhead, so that, in its light, their shadows changed size.

If my grope seem to have no grasp in it, and, if a stationary balloon will, in half an hour, not cast a stationary shadow from the setting sun, we have to think of two triangular objects that accurately maintained positions in a line between sun and clouds, and at the same time approached and receded from clouds. Whatever it may have been, it's enough to make the devout make the sign of the crucible, or whatever the devotees of the Old Dominant do in the presence of a new correlate.

Vast, black thing poised like a crow over the moon.

It is our acceptance that these two shadows of Chisbury looked, from the moon, like vast things, black as crows, poised over the earth. It is our acceptance that two triangular luminosities and then two triangular patches, like vast black things, poised like crows over the moon, and, like the triangularities at Chisbury, have been seen upon, or over, the moon:

Scientific American, 46–49:

Two triangular, luminous appearances reported by several observers in Lebanon, Conn., evening of July 3, 1882, on the moon's upper limb. They disappeared and two dark triangular appearances that looked like notches were seen three minutes later upon the lower limb. They approached each other, met and instantly disappeared.

The merger here is notches that have at times been seen upon the moon's limb: thought to be cross sections of craters (*Monthly Notices, R.A.S.*, 37–432). But these appearances of July 3, 1882, were vast upon the moon—"seemed to be cutting off or obliterating nearly a quarter of its surface".

Something else that may have looked like a vast black crow poised over this earth from the moon:

Monthly Weather Review, 41–599:

Description of a shadow in the sky, of some unseen body, April 8, 1913, Fort Worth, Texas—supposed to have been cast by an unseen cloud—this patch of shade moved with the declining sun.

Rept. Brit. Assoc., 1854–410:

Account by two observers of a faint but distinctly triangular object, visible for six nights in the sky. It was observed from two stations that were not far apart. But the parallax was considerable. Whatever it was, it was, acceptably, relatively close to this earth.

I should say that relatively to phenomena of light we are in confusion as great as some of the discords that orthodoxy is in relatively to light. Broadly and intermediatistically, our position is:

That light is not really and necessarily light—any more than is anything else really and necessarily anything—but an interpretation of a mode of force, as I suppose we have to call it, as light. At sea level, the earth's atmosphere interprets sunlight as red or orange or yellow. High up on mountains the sun is blue. Very high up on mountains the zenith is black. Or it is orthodoxy to say that in inter-planetary space, where there is no air, there is no light. So then the sun and comets are black, but the earth's atmo-sphere, or, rather, dust particles in it, interpret radiations from these black objects as light.

We look up at the moon.

The jet-black moon is so silvery white.

I have about fifty notes indicating that the moon has atmosphere: nevertheless most astronomers hold out that the moon has no atmosphere. They have to: the theory of eclipses would not work out otherwise. So, arguing in con-ventional terms, the moon is black. Rather astonish-ing—explorers upon the moon—stumbling and groping in intense darkness—with telescopes powerful enough, we could see them stumbling and groping in brilliant light.

Or, just because of familiarity, it is not now obvious to us how the preposterousness of the old system must have seemed to the correlated of the system preceding it.

Ye jet-black silvery moon.

Altogether, then, it may be conceivable that there are phenomena of force that are interpretable as light as far down as the clouds, but not in denser strata of air, or just the opposite of familiar interpretations.

I now have some notes upon an occurrence that suggests a force not interpreted by air as light, but interpreted, or reflected by the ground as light. I think of something that, for a week, was suspended over London: of an emanation that was not interpreted as light until it reached the ground.

Lancet, June 1, 1867:

That every night for a week, a light had appeared in Woburn Square, London, upon the grass of a small park, enclosed by railings. Crowds gathering—police called out "for the special service of maintaining order and making the populace move on".

The Editor of the *Lancet* went to the Square. He says that he saw nothing but a patch of light falling upon an arbor at the northeast corner of the enclosure. Seems to me that that was interesting enough.

In this Editor we have a companion for Mr. Symons and Dr. Gray. He suggests that the light came from a street lamp—does not say that he could trace it to any such origin himself—but recommends that police investigate neigh-bouring street lamps.

I'd not say that such a commonplace as light from a street lamp would not attract and excite and deceive great crowds for a week—but I do accept that any policeman who was called upon for extra work would have need nobody's suggestion to settle that point the very first thing.

Or that something in the sky hung suspended over a London Square for a week.

Chapter 21

Knowledge, Dec. 28, 1883:

"Seeing so many meteorological phenomena in your excellent paper, *Knowledge*, I am tempted to ask for an explanation of the following, which I saw when on board the British India Company's steamer *Patna*, while on a voyage up the Persian Gulf. In May, 1880, on a dark night, about 11.30 P.M., there suddenly appeared on each side of the ship an enormous luminous wheel, whirling around, the spokes of which seem to brush the ship along. The spokes would be 200 or 300 yards long, and resembled the birch rods of the dames' schools. Each wheel contained about sixteen spokes and, although the wheels must have been some 500 or 600 yards in diameter, the spokes could be distinctly seen all the way round. The phosphorescent gleam seemed to glide along flat on the surface of the sea, no light being visible in the air above the water. The appearance of the spokes could be almost exactly represented by standing in a boat and flashing a bull's eye lantern horizontally along the surface of the water, round and round. I may mention that the phenomena was also seen by Captain Avern, of the *Patna*, and Mr. Manning, third officer.

"Lee Fore Brace.

"P.S.—The wheels advanced along with the ship for about twenty minutes.—L.F.B."

Knowledge, Jan. 11, 1884:

Letter from "A. Mc. D.":

That "Lee Fore Brace", "who sees 'so many meteorological phenomena in your excellent paper', should have signed himself 'The Modern Ezekiel', for his vision of wheels is quite wonderful as the prophet's". The writer then takes up the measurements that were given, and calculates a velocity at the circumference of a wheel, of about 166 yards per second, apparently considering that especially incredible, He then says: "From the nom de plume he assumes, it might be inferred that your correspondent is in the habit of 'sailing close to the wind'." He asks permission to suggest an explanation of his own. It is that before 11.30

P.M. there had been numerous accidents to the "main brace", and that it had required splicing so often that almost any ray of light would have taken on a rotary motion.

In *Knowledge*, Jan. 25, 1884, Mr. "Brace" answers and signs himself "J. W. Robertson":

"I don't suppose A. Mc. D. means any harm, but I do think it's rather unjust to say a man is drunk because he sees something out of the common. If there's one thing I pride myself upon, it's being able to say that never in my life have I indulged in anything stronger than water." From this curiosity of pride, he goes on to say that he had not intended to be exact, but to give his impressions of dimensions and velocity. He ends amiably: "However, 'no offence taken, where I suppose none is meant'."

To this letter Mr. Proctor adds a note, apologizing for the publication of "A. Mc. D's." letter, which had come about by a misunderstood instruction. Then Mr. Proctor wrote disagreeable letters himself, about other persons—what else would you expect in a quasi-existence?

The obvious explanation of this phenomena is that, under the surface of the sea, in the Persian Gulf, was a vast luminous wheel: that it was the light from its submerged spokes that Mr. Robertson saw, shining upward. It seems clear that this light did shine upward from origin below the surface of the sea. But at first it is not so clear how vast luminous wheels, each the size of a village, ever got under the surface of the Persian Gulf: also there be some misunderstanding as to what they were doing there.

A deep-sea fish, and its adaptation to a dense medium—

That, at least in some regions aloft, there is a medium dense even to gelatinousness—

A deep-sea fish, brought to the surface of the ocean: in a relatively attenuated medium, it disintegrates—

Super-constructions adapted to a dense medium in inter-planetary space—sometimes, by stresses of various kinds, they are driven into this earth's thin atmosphere—

Later we shall have data to support just this: that things entering this earth's atmosphere disintegrate and shine with a light that is not the light of incandescence: shine brilliantly even if cold.

Vast wheel-like super-constructions—they enter this earth's atmosphere, and, threatened with disintegration, plunge for relief into an ocean, or into a denser medium.

Of course the requirements now facing us are:

Not only data of vast wheel-like super-constructions that have relieved their distresses in the ocean, but data of enormous wheels that have been seen in the air, or entering the ocean, or rising from the ocean and continuing their voyages.

Very largely we shall concern ourselves with enormous fiery objects that have either plunged into the ocean or risen from the ocean. Our acceptance is that, though disruption may intensify into incandescence, apart from disruption and its probable fieriness, things that enter this earth's atmosphere have a cold light which would not, like light from molten matter, be instantly quenched by water. Also it seems that a revolving wheel would, from a distance, look like a globe; that a revolving wheel, seen relatively close by, looks like a wheel in few aspects. The mergers of ball-lighting and meteorites are not resistances to us: our data are of enormous bodies.

So we shall interpret—and what does it matter?

Our attitude throughout this book:

That here are extraordinary data—that they never would be exhumed, and never would be massed together, unless—

Here are the data:

Our first datum is of something that was seen to enter an ocean. It's from the puritanic publication, *Science*, which has yielded us little material, or which, like most puritans, does not go upon a spree very often. Whatever the thing could have been, my impression is of tremendousness, or of bulk many times that of all meteorites in all museums combined: also of relative slowness, or of long warning of approach. The story, in *Science*, 5–242, is from an account sent to the Hydrographic Office, at Washington, from the branch office, at San Francisco: That, at midnight, Feb. 24, 1885, Lat. 37° N., and Long. 170° E., or somewhere between Yokohama and Victoria, the captain of the bark *Innerwich* was roused by his mate, who had seen something unusual in the sky. This must have taken appreciable time. The captain went on deck and saw the sky turning fiery red. "All at once, a large mass of fire appeared over the vessel, completely blinding the spectators." The fiery mass fell into the sea. Its size may be judged by the volume of water cast

283

up by it, said to have rushed toward the vessel with a noise that was "deafening". The bark was struck flat aback, and "a roaring, white sea passed ahead". "The master, an old experienced mariner, declared that the awfulness of the sight was beyond description."

In *Nature*, 37–38, and *L'Astronomie*, 1887–76, we are told that an object, described as "a large ball of fire", was seen to rise from the sea, near Cape Race. We are told that it rose to a height of fifty feet, and then advanced close to the ship, then moving away, remaining visible about five minutes. The supposition in *Nature* is that it was "ball lightning", but Flammarion, *Thunder and Lightning*, p. 68, says that it was enormous. Details in the American *Meteorological Journal*, 6–443—Nov. 12, 1887—British steamer *Siberian*—that the object had moved "against the wind" before retreating—that Captain Moore said that at about the same place he had seen such appearances before.

Report of the British Association, 1886–30:

That, upon June 18, 1845, according to the *Malta Times*, from the brig *Victoria*, about 900 miles east of Adalia, Asia Minor (36° 40′ 56″ N. Lat.: 13° 44′ 36″ E. Long.), three luminous bodies were seen to issue from the sea, at about half a mile from the vessel. They were visible about ten minutes.

The story was never investigated, but other accounts that seem acceptably to be other observations upon this same sensational spectacle came in, as if of their own accord, and were published by Prof. Baden-Powell. One is a letter from a correspondent at Mt. Lebanon. He describes only two luminous bodies. Apparently they were five times the size of the moon: each had appendages, or they were connected by parts that are described as "sail-like or steamer-like", looking like "large flags blown out by a gentle breeze". The important point here is not only suggestion of structure, but duration. The duration of meteors is a few seconds; duration of fifteen seconds is remarkable, but I think there are records up to half a minute. This object, if it were all one object, was visible at Mt. Lebanon about one hour. An interesting circumstance is that the appendages did not look like trains of meteors, which shine by their own light, but "seemed to shine by light from the main bodies".

About 900 miles west of the position of the *Victoria* is the town of Adalia, Asia Minor. At about the time of the observation reported by the captain of the *Victoria*, the Rev. F. Hawlett, F.R.A.S., was in Adalia. He, too, saw this spectacle, and sent an account to Prof. Baden-Powell. In his view it was a body that appeared and then broke up. He places duration at twenty minutes to half an hour.

In the *Report of the British Association*, 1860–82, the phenomenon was reported from Syria and Malta, as two very large bodies "nearly joined".

Rept. Brit. Assoc., 1860–77:

That, at Cherbourg, France, Jan. 12, 1836, was seen a luminous body, seemingly, two-thirds the size of the moon. It seemed to rotate on an axis. Central to it there seemed to be a dark cavity.

For other accounts, all indefinite, but distortable into data of wheel-like objects in the sky, see *Nature*, 22–617; London *Times*, Oct. 15, 1859; *Nature*, 21–225; *Monthly Weather Review*, 1883–264.

L'Astronomie, 1894–157:

That, upon the morning of Dec. 20, 1893, an appearance in the sky was seen by many persons in Virginia, North Carolina, and South Carolina. A luminous body passed overhead, from west to east, until at about 15 degrees in the eastern horizon, it appeared to stand still for fifteen or twenty minutes. According to some descriptions it was the size of a table. To some observers it looked like an enormous wheel. The light was a brilliant white. Acceptably it was not an optical illusion—the noise of its passage through the air was heard. Having been stationary, or having seemed to stand still fifteen or twenty minutes, it disappeared, or exploded. No sound of explosion was heard.

Vast wheel-like constructions. They're especially adapted to roll through a gelatinous medium from planet to planet. Sometimes, because of miscalculations, or because of stresses of various kinds, they enter this earth's atmosphere. They're likely to explode. They have to submerge in the sea. They stay in the sea a while, revolving with relative leisureliness, until relieved, and then emerge, sometimes close to vessels. Seamen tell of what they see: their reports are interred in scientific morgues. I should say that the general route of these constructions is along latitudes not far from the latitudes of the Persian Gulf.

Journal of the Royal Meteorological Society, 28–29:

That, upon April 4, 1901, about 8.30, in the Persian

Gulf, Captain Hoseason, of the steamship *Kilwa*, according to a paper read before the Society by Captain Hoseason, was sailing in a sea in which there was no phosphorescence—"there being no phosphorescence in the water".

I suppose I'll have to repeat that:

". . . there being no phosphorescence in the water".

Vast shafts of light—though the captain uses the word "ripples"—suddenly appeared. Shaft followed shaft, upon the surface of the sea. But it was only a faint light, and, in about fifteen minutes, died out: having appeared suddenly, having died out gradually. The shafts revolved at a velocity of about 60 miles an hour.

Phosphorescent jellyfish correlate with the Old Dominant: in one of the most heroic compositions of disregards in our experience, it was agreed, in the discussion of Capt. Hoseason's paper, that the phenomenon was probably pulsations of long strings of jellyfish.

Nature, 21–410:

Reprint of a letter from R. E. Harris, Commander of the A.H.N. Co.'s steamship *Shahjehan*, to the Calcutta *Englishman*, Jan. 21, 1880:

That upon the 5th of June, 1880, off the coast of Malabar, at 10 P.M., water calm, sky cloudless, he had seen something that was so foreign to anything that he had ever seen before, that he had stopped his ship. He saw what he describes as waves of brilliant light, with spaces between. Upon the water were floating patches of a substance that was not identified. Thinking in terms of the conventional explanation of all phosphorescence at sea, the captain at first suspects this substance. However, he gives his opinion that it did no illuminating but was, with the rest of the sea, illuminated by tremendous shafts of light. Whether it was a thick and oily discharge from the engine of a submerged construction or not, I think that I shall have to accept this substance as a concomitant, because of another note. "As wave succeeded wave, one of the most grand and brilliant, yet solemn, spectacles that one could think of, was here witnessed."

Jour. Roy. Met. Soc., 32–280:

Extract from a letter from Mr. Douglas Carnegie, Blackheath, England. Date some time in 1906—

"This last voyage we witnessed a weird and most extraordinary electric display." In the Gulf of Oman, he saw a bank of apparently quiescent phosphorescence: but, when within twenty yards of it, "shafts of brilliant light came sweeping across the ship's bows at a prodigious speed, which might be put down as anything between 60 and 200 miles an hour". "These light bars were about 20 feet apart and most regular." As to phosphorescence—"I collected a bucketful of water, and examined it under the microscope, but could not detect anything abnormal." That the shafts of light came up from something beneath the surface—"They first struck us on our broadside, and I noticed that an intervening ship had no effect on the light beams: they started away from the lee side of the ship, just as if they had travelled right through it."

The Gulf of Oman is at the entrance to the Persian Gulf.

Jour. Roy. Met. Soc., 33–294:

Extract from a letter by Mr. S. C. Patterson, second officer of the P. and O. steamship *Delta*: a spectacle which the *Journal* continues to call phosphorescent:

Malacca Strait, 2 A.M., March 14, 1907:

"... shafts which seemed to move round a centre—like the spokes of a wheel—and appeared to be about 300 yards long. The phenomenon lasted about half an hour, during which time the ship had travelled six or seven miles. It stopped suddenly."

L'Astronomie, 1891-312:

A correspondent writes that, in October, 1891, in the China Sea, he had seen shafts or lances of light that had had the appearance of rays of a searchlight, and that had moved like such rays.

Nature, 20–291:

Report to the Admiralty by Capt. Evans, the Hydrographer of the British Navy:

That Commander J. E. Pringle, of H.M.S. *Vulture*, had reported that, at Lat. 26° 26′ N., and Long. 53° 11′ E.—in the Persian Gulf—May 15, 1879, he had noticed luminous waves or pulsations, in the water, moving at great speed. This time we have a definite datum upon origin somewhere below the surface. It is said that these waves of light passed under the *Vulture*. "On looking toward the east, the appearance was that of a revolving wheel with a centre on that bearing, and those spokes were illuminated, and, looking toward the west, a similar wheel appeared to be revolving, but in the opposite direction." Or finally as to submergence—"These waves of light extended from surface well under the water." It is Commander Pringle's opinion that the shafts constituted one wheel, and that doubling was an illusion. He judges the shafts to have been about 25 feet broad, and the spaces about 100. Velocity about 84 miles an hour. Duration about 35 minutes. Time 9.40 P.M. Before and after this display the ship had passed through patches of floating substance described as "oily-looking fish spawn".

Upon page 428 of this number of *Nature*, E. L. Moss says that, in April, 1875, when upon H.M.S. *Bulldog*, a few miles north of Vera Cruz, he had seen a series of swift lines of light. He had dipped up some of the water, finding in it animalcule, which would, however, not account for phenomena of geometric formation and high velocity. If he means Vera Cruz, Mexico, this is the only substance we have out of oriental waters.

Scientific American, 106–51:

That, in the *Nautical Meteorological Annual*, published by the Danish Meteorological Institute, appears a report upon a "singular phenomenon" that was seen by Capt. Gabe, of the Danish East Asiatic Co.'s steamship *Bintang*. At 3 A.M., June 10, 1909, while sailing through the Straits of Malacca, Captain Gabe saw a vast revolving wheel of light, flat upon the water—"long arms issuing from the centre around which the whole system appeared to rotate". So vast was the appearance that only half of it could be seen at a time, the centre lying near the horizon. This display lasted about fifteen minutes. Heretofore we have not been clear upon the important point that forward motions of these wheels do not synchronize with a vessel's motions, and freaks of disregard, or, rather, commonplaces of disregard, might attempt to assimilate with lights of a vessel. This time we are told that the vast wheel moved forward, decreasing in brilliancy, and also in speed of rotation, disappearing when the centre was right ahead of the vessel—or my own interpretation would be that the source of light was submerging deeper and deeper and slowing down because meeting more and more resistance.

The Danish Meteorological Institute reports another instance:

That, when Capt. Breyer, of the Dutch steamer *Valentijn*, was in the South China Sea, midnight, Aug. 12, 1910, he saw a rotation of flashes. "It looked like a horizontal wheel, turning rapidly." This time it is said that the appearance was above water. "The phenomenon was observed by the captain, the first and second mates, and the first engineer, and upon all of them it made a somewhat uncomfortable impression."

In general, if our expression be not immediately acceptable, we recommend to rival interpreters that they consider the localization—with one exception—of this phenomenon, to the Indian Ocean and adjacent waters, or Persian Gulf on one side and China Sea on the other side. Though we're Intermediatists, the call of attempted Positivism, in the aspect of Completeness, is irresistible. We have expressed that from few aspects would wheels of fire in the air look like wheels of fire, but, if we can get it, we must have observation upon vast luminous wheels, not interpretable as optical illusions, but enormous, substantial things that have smashed down material resistances, and have been seen to plunge into the ocean:

Athenaeum, 1848–833:

That at the meeting of the British Association, 1848, Sir W. S. Harris said that he had recorded an account sent to him of a vessel toward which had whirled "two wheels of fire, which the men described as rolling millstones of fire". "When they came near, an awful crash took place: the topmasts were shivered to pieces." It is said that there was a strong sulphurous odour.

Chapter 22

Journal of the Royal Meteorological Society, 1–157:

Extract from the log of the bark *Lady of the Lake*, by Capt. F. W. Banner:

Communicated by R. H. Scott, F.R.S.:

That upon the 22nd of March, 1870, at Lat. 5° 47′ N., Long. 27° 52′ W., the sailors of the *Lady of the Lake* saw a remarkable object, or "cloud", in the sky. They reported to the captain.

According to Capt. Banner, it was a cloud of circular form, with an included semicircle divided into four parts, the central dividing shaft beginning at the centre of the circle and extending far outward, and then curving backward.

Geometricity and complexity and stability of form: and the small likelihood of a cloud maintaining such diversity of features, to say nothing of appearance of organic form.

The thing travelled from a point at about 20 degrees above the horizon to a point about 80 degrees above. Then it settled down to the northwest, having appeared from the south, southeast.

Light grey in colour, or it was cloud-colour.

"It was much lower than the other clouds."

And this datum stands out:

That, whatever it may have been, it travelled against the wind.

"It came up obliquely against the wind, and finally settled down right in the wind's eye."

For half an hour this form was visible. When it did finally disappear that was not because it disintegrated like a cloud, but because it was lost to sight in the evening darkness.

Capt. Banner draws the following diagram:

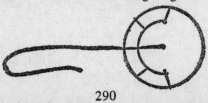

Chapter 23

Text-books tell us that the Dhurmsalla meteorites were picked up "soon", or "within half an hour". Given a little time the conventionalists may argue that these stones were hot when they fell, but that their interior coldness had overcome the molten state of their surfaces.

According to the Deputy Commissioner of Dhurmsalla, these stones had been picked up "immediately" by passing coolies.

These stones were so cold that they numbed the fingers. But they had fallen with a great light. It is described as "a flame of fire about two feet in depth and nine feet in length". Acceptably this light was not the light of molten matter.

In this chapter we are very intermediastic—and unsatisfactory. To the intermediatist there is but one answer to all questions:

Sometimes and sometimes not.

Another form of this "intermediatist solution" of all problems is:

Yes and no.

Everything that is, also isn't.

A positivist attempts to formulate: so does the intermediatist, but with less rigorousness: he accepts but also denies: he may seem to accept in one respect and deny in some other respect, but no real line can be drawn between any two aspects of anything. The intermediatist accepts that which seems to correlate with something that he has accepted as a dominant. The positivist correlates with a belief.

In the Dhurmsalla meteorites we have support for our expression that things entering this earth's atmosphere sometimes shine with a light that is not the light of incandescence—or so we account, or offer an expression upon, "thunderstones," or carved stones that have fallen luminously to this earth, in streaks that have looked like strokes of lightning—but we accept, also, that some things that have entered this earth's atmosphere, disintegrate with the intensity of flame and molten matter—but some things, we accept enter this earth's atmosphere and collapse non-luminously, quite like deep-sea fishes brought to the surface of the ocean. Whatever agreement we have is an indication that somewhere aloft there is a medium denser than this earth's atmosphere. I suppose our stronghold is in that such is not popular belief—

Or the rhythm of all phenomena:

Air dense at sea level upon this earth—less and less dense as one ascends—then denser and denser. A good many bothersome questions arise—

Our attitude:

Here are the data:

Luminous rains sometimes fall (*Nature*, March 9, 1882; *Nature*, 25–437). This is light that is not the light of incandescence, but no one can say that these occasional, or rare, rains come from this earth's externality. We simply note cold light of falling bodies. For luminous rain, snow, and dust, see Hartwig, *Aerial World*, p. 319. As to luminous clouds, we have more nearly definite observations and opinions: they mark transition between the Old Dominant and the New Dominant. We have already noted the transition in Prof. Schwedoff's theory of external origin of some hailstones—and the implications that, to a former generation, seemed so preposterous—"droll" was the word—that there are in inter-planetary regions volumes of water—whether they have fishes and frogs in them or not. Now our acceptance is that clouds sometimes come from external regions, having had origin from super-geographical lakes and oceans that we shall not attempt to chart, just at present—only suggesting to enterprising aviators—and we note that we put it all up to them, and show no inclination to go Columbusing on our own account—that they take bathing suits, or, rather, deep-sea diving-suits along. So then that some clouds from inter-planetary oceans—of the Super-Sargasso Sea—if we still accept the Super-Sargasso Sea—and shine, upon entering this earth's atmosphere. In *Himmel und Erde*, February, 1889—a phenomenon of transition of thirty years ago—Herr O. Jesse, in his observations upon luminous night clouds, notes that great height of them, and drolly or sensibly suggests that some of them may have come from regions external to this earth. I suppose he means only from other planets. But it's a very droll and sensible idea either way.

In general I am accounting for a great deal of this earth's isolation: that it is relatively isolated by circumstances that are similar to the circumstances that make for relative isolation of the bottom of the ocean—except that there is a clumsiness of analogy now. To call ourselves deep-sea fishes has been convenient, but, in a quasi-existence, there is no convenience that will not sooner or later turn awkward—so, if there be denser regions aloft, these regions should be regarded as analogues of far-submerged oceanic regions, and things coming to this earth would be like things rising to an attenuated medium—and exploding—sometimes incandescently, like deep-sea fishes brought to the surface—altogether conditions of inhospitality. I have a suspicion that, in their depths, deep-sea fishes are not luminous. If they are, Darwinism is mere Jesuitism, in attempting to corrrelate them. Such advertising would so attract attention that all advantages would be more than offset. Darwinism is largely a doctrine of concealment: here we have brazen proclamation—if accepted. Fishes in the Mammoth Cave need no light to see by. We might have an expression that deep-sea fishes turn luminous upon entering a less dense medium—but models in the American Museum of Natural History: specialized organs of luminosity upon these models. Of course we do remember that awfully convincing "dodo", and some of our sophistications we trace to him—at any rate disruption is regarded as a phenomenon of coming from a dense to a less dense medium.

An account by M. Acharius, in the *Transactions of the Swedish Academy of Sciences*, 1808–215, translated for the *North American Review*, 3–319:

That M. Acharius, having heard of "an extraordinary and probably hitherto unseen phenomenon", reported from near the town of Skeninge, Sweden, investigated:

That, upon the 16th May, 1808, at about 4 P.M., the sun suddenly turned dull brick-red. At the same time there appeared upon the western horizon, a great number of round bodies, dark brown, and seemingly the size of a hat crown. They passed overhead and disappeared in the eastern horizon. Tremendous procession. It lasted two hours. Occasionally one fell to the ground. When the place of fall was examined, there was found a film, which soon dried and vanished. Often, when approaching the sun, these bodies seemed to link together, or were then seen to be linked together, in groups not exceeding eight, and, under the sun, they were seen to have tails three or four fathoms long. Away from the sun the tails were invisible. Whatever their substance may have been it is described as gelatinous—"soapy and jellied".

I place this datum here for several reasons. It would have been a good climax to our expression upon hordes of small bodies that, in our acceptance, were not seeds, nor birds, nor ice-crystals: but the tendency would have been to jump to the homogeneous conclusion that all our data in that expression related to this one kind of phenomena, whereas we conceive of infinite heterogeneity of the external: of crusaders and rabbles and emigrants and tourists and dragons and things like gelatinous hat crowns. Or that all things, here, upon this earth, that flock together, are not necessarily sheep, Presbyterians, gangsters, or porpoises. The datum is important to us here as indication of disruption in this earth's atmosphere—dangers in entering this earth's atmosphere.

I think myself, that thousands of objects have been seen to fall from aloft, and have exploded luminously, and have been called "ball lightning".

"As to what ball lightning is, we have not yet begun to make intelligent guesses." (*Monthly Weather Review*, 34–17.)

In general, it seems to me that when we encounter the opposition "ball lightning" we should pay little attention, but confine ourselves to guesses that are at least intelligent, that stand phantom-like in our way. We note here that in some of our acceptances upon intelligence we should more clearly have pointed out that they were upon the intelligent as opposed to the instinctive. In the *Monthly Weather Review*, 33–409, there is an account of "ball lightning", that struck a tree. It made a dent such as a falling object would make. Some other time I shall collect instances of "ball lightning" to express that they are instances of objects that have fallen from the sky, luminously, exploding terrifically. So bewildered is the old orthodoxy by these phenomena that many scientists have either denied "ball lightning" or have considered it very doubtful. I refer to Dr. Sestier's list of one hundred and fifty instances, which he considered authentic.

In accord with our disaccord is an instance related in the *Monthly Weather Review*, March, 1887—something that fell luminously from the sky, accompanied by something that was not so affected, or that was dark:

That according to Capt. C. D. Sweet, of the Dutch bark, *J.P.A.*, upon March 19, 1887, 37° 39′ N., 57° 00′ W., he encountered a severe storm. He saw two objects in the air above the ship. One was luminous, and might be explained in several ways, but the other was dark. One or both fell into the sea, with a roar and the casting up of billows. It is our acceptance that these things had entered this earth's atmosphere, having first crashed through a field of ice—"immediately afterward lumps of ice fell".

One of the most astonishing of the phenomena of "ball lightning" is a phenomena of many meteorites: violence of explosion out of all proportion to size and velocity. We accept that the icy meteorites of Dhurmsalla could have fallen with no great velocity, but the sound from them was tremendous. The soft substance that fell at the Cape of Good Hope was carbonaceous, but was unburned, or had fallen with velocity insufficient to ignite it. The tremendous report that it made was heard over an area more than seventy miles in diameter.

That some hailstones have been formed in a dense medium, and violently disintegrate in this earth's relatively thin atmosphere:

Nature, 88–350:

Large hailstones noted at the University of Missouri, Nov. 11, 1911: they exploded with sounds like pistol shots. The writer says that he had noticed a similar phenomenon, eighteen years before, at Lexington, Kentucky. Hailstones that seemed to have been formed in a denser medium: when melted under water they gave out bubbles larger than their central air spaces. (*Monthly Weather Review*, 33–445.)

Our acceptance is that many objects have fallen from the sky, but that many of them have disintegrated violently. This acceptance will co-ordinate with data still to come, but, also, we make it easy for ourselves in our expressions upon super-constructions, if we're asked why, from thinkable wrecks of them, girders, plates, or parts recognizably of manufactured metal have not fallen from the sky.

However, as to composition, we have not this refuge, so it is our expression that there have been reported instances of the fall of manufactured metal from the sky.

The meteorite of Rutherford, North Carolina, is of artificial material: mass of pig iron. It is said to be fraudulent. (*Amer. Jour. Sci.*, 2–34–298.)

The object that was said to have fallen at Marblehead, Mass., in 1858, is described in the *Amer. Jour. Sci.*, 2–34–135, as "a furnace product, formed in smelting copper ores, or iron ores containing copper". It is said to be fraudulent.

According to Ehrenburg, the substance reported by Capt. Callam to have fallen upon his vessel, near Java, "offered complete resemblance to the residue resulting from combustion of a steel wire in a flask of oxygen". (Zurcher, *Meteors*, p. 239.) *Nature*, Nov. 21, 1878, publishes a notice that, according to the *Yuma Sentinel*, a meteorite that "resembles steel" had been found in the Mohave Desert. In *Nature*, Feb. 15, 1894, we read that one of the meteorites brought to the United States by Peary, from Greenland, is of tempered steel. The opinion is that meteoric iron had fallen in water or snow, quickly cooling and hardening. This does not apply to composition. Nov. 5, 1898, *Nature* publishes a notice of a paper by Prof. Berwerth, of Vienna, upon "the close connection between meteoric iron and steel-works' steel".

At the meeting of Nov. 24, 1906, of the Essex Field Club, was exhibited a piece of metal said to have fallen from the sky, Oct. 9, 1906, at Braintree. According to the *Essex Naturalist*, Dr. Fletcher, of the British Museum, had declared this metal to be smelted iron—"so that the mystery of its reported 'fall' remained unexplained".

Chapter 24

We shall have an outcry of silences. If a single instance of
anything be disregarded by a System—our own attitude is
that a single instance is a powerless thing. Of course our
own method of agreement of many instances is not a real
method. In Continuity, all things must have resemblances
with all other things. Anything has any quasi-identity you
please. Some time ago conscription was assimilated with
either autocracy or democracy with equal facility. Note the
need for a dominant to correlate to. Scarcely anybody said
simply that we must have conscription: but that we must
have conscription, which correlates with democracy, which
was taken as a base, or something basically desirable. Of
course between autocracy and democracy nothing but false
demarcation can be drawn. So I can conceive of no subject
upon which there should be such poverty as a single
instance if anything one pleases can be whipped into line.
However, we shall try to be more nearly real than the
Darwinites who advance concealing colouration as Darwinism, and then drag in proclaiming luminosity, too, as Darwinism. I think the Darwinites had better come in with us as
to the deep-sea fishes—and be sorry later, I suppose. It will
be amazing or negligible to read all the instances now to
come of things that have been seen in the sky, and to think
that all have been disregarded. My own opinion is that it is
not possible, or very easy, to disregard them, now that they
have been brought together—but that, if prior to about this
time we had attempted such an assemblage, the Old
Dominant would have withered our typewriter—as it is the
letter "e" has gone back on us, and the "s" is tempera-
mental.

"Most extraordinary and singular phenomenon", North
Wales, Aug. 26, 1894; a disc from which was projected an
orange-coloured body that looked like "an elongated
flatfish", reported by Admiral Ommanney (*Nature*,
50–524); disc from which projected a hook-like form,
India, about 1838; diagram of it given; disc about size of
the moon, but brighter than the moon; visible about twenty

minutes; by G. Pettit, in Prof. Baden-Powell's Catalogue (*Rept. Brit. Assoc.*, 1849); very brilliant hook-like form, seen in the sky at Poland, Trumbell Co., Ohio, during the stream of meteors, of 1833; visible more than an hour: large luminous body, almost stationary "for a time"; shaped like a square table; Niagara Falls, Nov. 13, 1833 (*Amer. Jour. Sci.*, 1-25-391); something described as a bright white cloud, at night, Nov. 3, 1886, at Hamar, Norway; from it were emitted brilliant rays of light; drifted across the sky; "retained throughout its original form" (*Nature*, Dec. 16, 1886-158); thing with an oval nucleus, and streamers with dark bands and lines very suggestive of structure; New Zealand, May 4, 1888 (*Nature*, 42-402); luminous object, size of full moon, visible an hour and a half, Chili, Nov. 5, 1883 (*Comptes Rendus*, 103-682); bright object near sun, Dec. 21, 1882 (*Knowledge*, 3-13); light that looked like a great flame, far out at sea, off Ryook Phyoo, Dec. 2, 1845 (*London Roy. Soc. Proc.*, 5-627); something like a gigantic trumpet, suspended vertical, oscillating gently, visible five or six minutes, length estimated at 425 feet, at Oaxaca, Mexico, July 6, 1874 (*Sci. Am. Sup.*, 6-2365); two luminous bodies, seemingly united, visible five or six minutes, June 3, 1898 (*La Nature*, 1898-1-127); thing with a tail, crossing the moon, transit half a minute. Sept. 26, 1870 (*London Times*, Sept. 30, 1870); object four or five times size of moon, moving slowly across sky, Nov. 1, 1885, near Adrianople (*L'Astronomie*, 1886-309); large body, coloured red, moving slowly visible 15 minutes by Coggia, Marseilles, Aug. 1, 1871 (*Chem. News*, 24-193); details of this observation, and similar observation by Guillemin, and other instances by de Fonville (*Comptes Rendus*, 73-297, 755); thing that was large and that was stationary twice in seven minutes, Oxford, Nov. 19, 1847; listed by Lowe (*Rec. Sci.*, 1-136); greyish object that looked to be about three and a half feet long, rapidly approaching the earth at Saarbruck, April 1, 1826; sound like thunder; object expanding like a sheet (*Amer. Jour. Sci.*, 1-26-133; *Quar. Jour. Roy. Inst.*, 24-488); report by an astronomer, N. S. Drayton, upon an object duration of which seemed to him extraordinary; duration three-quarters of a minute, Jersey City, July 6, 1882 (*Sci. Amer.*, 47-53); object like a comet, but with proper motion of 10 degrees an hour; visible one hour; reported by Purine and

Glancy from the Cordoba Observatory, Argentina, March 14, 1916 (*Sci. Amer.*, 115–493); something like a signal light, reported by Glaisher, Oct. 4, 1844; bright as Jupiter, "sending out quick flickering waves of light". (*Year Book of Facts*, 1845–278.)

I think that with the object known as Eddie's "comet" passes away the last of our susceptibility to the common fallacy of personifying. It is one of the most deep-rooted of positivist illusions—that people are persons. We have been guilty too often of spleens and spite and ridicules against astronomers, as if they were persons, or final unities, individuals, completenesses, or selves—instead of inde-terminate parts. But, so long as we remain in quasi-existence, we can cast out illusion only with some other illusion, though the other illusion may approximate higher to reality. So we personify no more—but we super-personify. We now take into full acceptance our expression that Development is an Autocracy of Successive Domin-ants—which are not final—but which approximate higher to individuality of self-ness, than do the human tropisms that irresponsibly correlate to them.

Eddie reported a celestial object, from the Observatory at Grahamstown, South Africa. It was in 1890. The New Dominant was only heir presumptive then, or heir apparent but not obvious. The thing that Eddie reported might as well have been reported by a night watchman, who had looked up through an unplaced sewer pipe.

It did not correlate.

The thing was not admitted to *Monthly Notices*, I think myself that if the Editor had attempted to let it in—earth-quake—or a mysterious fire in his publishing house.

The Dominants are jealous gods.

In *Nature*, presumably a vassal of the new god, though of course also plausibly rendering homage to the old, is reported a comet-like body, of Oct. 27, 1890, observed at Grahamstown, by Eddie. It may have looked comet-like, but it moved 100 degrees while visible, or one hundred degrees in three-quarters of an hour. See *Nature*, 43–89, 90.

In *Nature*, 44–519, Prof. Copeland describes a similar appearance that he had seen, Sept. 10, 1891. Dreyer says (*Nature*, 44–541) that he had seen this object at the Armagh Observatory. He likens it to the object that was reported by Eddie. It was seen by Dr. Alexander Graham Bell, Sept. 11, 1891, in Nova Scotia.

But the Old Dominant was a jealous god.

So there were different observations upon something that was seen in November, 1883. These observations were Philistines in 1883. In the *Amer. Met. Jour.*, 1–110, a correspondent reports having seen an object like a comet, with two tails, one up and one down, Nov. 10 or 12, 1883. Very likely this phenomenon should be placed in our expression upon torpedo-shaped bodies that have been seen in the sky—our data upon dirigibles, or super-Zeppelins—but our attempted classifications are far from rigorous—or are mere groups. In the *Scientific American*, 50–40, a correspondent writes from Humacao, Porto Rico, that, Nov. 21, 1883, he and several other—persons—or persons, as it were—had seen a majestic appearance, like a comet. Visible three successive nights: disappeared then. The Editor says that he can offer no explanation. If accepted, this thing must have been close to the earth. If it had been a comet, it would have been seen widely, and the news would have been telegraphed over the world, says the Editor. Upon page 97 of this volume of the *Scientific American*, a correspondent writes that, at Sulphur Springs, Ohio, he had seen "a wonder in the sky", at about the same date. It was torpedo-shaped, or something with a nucleus, at each end of which was a tail. Again the Editor says that he can offer no explanation: that the object was not a comet. He associates it with the atmospheric effects general in 1883. But it will be our expression that, in England and Holland, a similar object was seen in November, 1882.

In the *Scientific American*, 40–294, is published a letter from Henry Harrison, of Jersey City, copied from the *New York Tribune*: that upon the evening of April 13, 1879, Mr. Harrison was searching for Brorsen's comet, when he saw an object that was moving so rapidly that it could not have been a comet. He called a friend to look, and his observation was confirmed. At two o'clock in the morning this object was still visible. In the *Scientific American Supplement*, 7–2885, Mr. Harrison disclaims sensationalism, which he seems to think unworthy, and gives technical details: he says that the object was seen by Mr. J. Spencer Devoe, of Manhattanville.

Chapter 25

"A formation having the shape of a dirigible." It was reported from Huntington, West Virginia (*Sci. Amer.*, 115-241). Luminous object that was seen July 19, 1916, at about 11 P.M. Observed through "rather powerful field glasses", it looked to be about two degrees long and half a degree wide. It gradually dimmed, disappeared, reappeared, and then faded out of sight. Another person—as we say: it would be too inconvenient to hold to our intermediatist recognitions—another person who observed this phenomenon suggested to the writer of the account that the object was a dirigible, but the writer says that faint stars could be seen behind it. This would seem really to oppose our notion of a dirigible visitor to this earth—except for the inconclusiveness of all things in a mode of seeming that is not final—or we suggest that behind some parts of the object, thing, construction, faint stars were seen. We find a slight discussion here. Prof. H. M. Russell thinks that the phenomenon was a detached cloud of aurora borealis. Upon page 369 of this volume of the *Scientific American* another correlator suggests that it was a light from a blast furnace— disregarding that, if there be blast furnaces in or near Huntington, their reflections would be commonplaces there.

We now have several observations upon cylindrical-shaped bodies that have appeared in this earth's atmosphere: cylindrical, but pointed at both ends, or torpedo-shaped. Some of the accounts are not very detailed, but out of the bits of description my own acceptance is that super-geographical routes are traversed by torpedo-shaped super-constructions that have occasionally visited, or that have occasionally been driven into this earth's atmosphere. From data, the acceptance is that upon entering this earth's atmosphere, these vessels have been so racked that had they not sailed away, disintegration would have occurred: that, before leaving this earth, they have, whether in attempted communication or not, or in mere wantonness or not, dropped objects, which did almost immediately violently disintegrate, or explode. Upon general principles we think that explosives have not been purposely dropped, but that parts have been racked off, and have fallen, exploding like the things called "ball lightning". May have been objects of stone or metal with inscriptions upon them, for all we know, at present. In all instances, estimates of dimensions are valueless, but ratios of dimensions are more acceptable. A thing said to have been six feet long may have been six hundred feet long; but shape is not so subject to the illusions of distance.

Nature, 40–415:

That, Aug. 5, 1889, during a violent storm, an object that looked to be about 15 inches long and 5 inches wide, fell, rather slowly at East Twickenham, England. It exploded. No substance from it was found.

L'Année Scientifique, 1864–54:

That, Oct. 10, 1864, M. Leverrier had sent to the Academy three letters from witnesses of a long luminous body, tapering at both ends, that had been seen in the sky.

In *Thunder and Lightning*, p. 87, Flammarion says that on Aug. 20, 1880, during a rather violent storm, M. A. Trécul, of the French Academy, saw a very brilliant yellowish-white body, apparently 35 to 40 centimetres long, and about 25 centimetres wide. Torpedo-shaped. Or a cylindrical body, "with slightly conical ends". It dropped something, and disappeared in the clouds. Whatever it may have been that was dropped, it fell vertically, like a heavy object, and left a luminous train. The scene of this occurrence may have been far from the observer. No sound was heard. For M. Trécul's account, see *Comptes Rendus*, 103–849.

Monthly Weather Review, 1907–310:

That, July 2, 1907, in the town of Burlington, Vermont, a terrific explosion had been heard throughout the city. A ball of light, or a luminous object, had been seen to fall from the sky – or from a torpedo-shaped thing, or construction, in the sky. No one had seen this thing that had exploded, fall from a larger body that was in the sky—but if we accept that at the same time there was a larger body in the sky—

My own acceptance is that a dirigible in the sky, or a construction that showed every sign of disrupting, had barely time to drop whatever it did drop—and to speed away to safety above.

The following story is told, in the *Review*, by the Bishop

John S. Michaud:

"I was standing on the corner of Church and College Streets, just in front of the Howard Bank. And facing east, engaged in conversation with Ex-Governor Woodbury and Mr. A. A. Buell, when, without the slightest indication, or warning, we were startled by what sounded like a most unusual and terrific explosion, evidently very nearby. Raising my eyes, and looking eastward along College Street, I observed a torpedo-shaped body, some 300 hundred feet away, stationary in appearance, and suspended in the air, about 50 feet above the tops of the buildings. In size it was about six feet long by eight inches in diameter, the shell, or covering, having a dark appearance, with here and there tongues of fire issuing from spots on the surface, resembling red-hot, unburnished copper. Although stationary when first noticed, this object soon began to move, rather slowly, and disappeared over Dolans Brothers' store, southward. As it moved, the covering seemed rupturing in places, and through these the intensely red flames issued."

Bishop Michaud attempts to correlate it with meteorological observations.

Because of the nearby view this is perhaps the most remarkable of the new correlates, but the correlate now coming is extraordinary because of the great number of recorded observations upon it. My own acceptance is that, upon Nov. 17, 1882, a vast dirigible crossed England, but by the definiteness-indefiniteness of all things quasi-real some observations upon it can be correlated with anything one pleases.

E. W. Maunder, invited by the Editors of the *Observatory* to write some reminiscences for the 500th number of their magazine, gives one that he says stands out (*Observatory*, 39–214). It is upon something that he terms "a strange celestial visitor". Maunder was at the Royal Observatory, Greenwich, Nov. 17, 1882, at night. There was an aurora, without features of special interest. In the midst of the aurora, a great circular disc of greenish light appeared and moved smoothly across the sky. But the circularity was evidently the effect of foreshortening. The thing passed above the moon, and was, by other observers, described as "cigar-shaped", "like a torpedo", "a spindle", "a shuttle". The idea of foreshortening is not mine: Maunder says this. He says: "Had the incident occurred a third of a century

303

later, beyond doubt everyone would have selected the same simile—it would have been 'just like a Zeppelin'." The duration was about two minutes. Colour said to have been the same as that of the aurora glow in the north. Nevertheless, Maunder says that this thing had no relation to auroral phenomena. "It appeared to be definite body." Motion too fast for a cloud, but "nothing could be more unlike the rush of a meteor". In the *Philosophical Magazine*, 5–15–318, J. Rand Capron, in a lengthy paper, alludes throughout to this phenomenon as an "auroral beam", but he lists many observations upon its "torpedo-shape", and one observation upon a "dark nucleus" in it—host of most confusing observations—estimates of height between 40 and 200 miles—observations in Holland and Belgium. We are told that according to Capron's spectroscopic observations the phenomenon was nothing but a beam of auroral light. In the *Observatory*, 6–192, in Maunder's contemporaneous account. He gives apparent approximate length and breadth at twenty-seven degrees and three degrees and a half. He gives other observations seeming to indicate structure—"remarkable dark marking down the centre".

In *Nature*, 27–84, Capron says that because of the moonlight he had been able to do little with the spectroscope.

Colour white, but aurora rosy (*Nature*, 27–87).

Bright stars seen through it, but not at the zenith, where it looked opaque. This is the only assertion of transparency (*Nature*, 27–87). Too slow for a meteor, but too fast for a cloud (*Nature*, 27–86). "Surface had a mottled appearance" *Nature*, 27–87). "Very definite in form, like a torpedo" (*Nature*, 27–100). "Probably a meteoric object" (Dr. Groneman, *Nature*, 27–296). Technical demonstration by Dr. Groneman, that it was a cloud of meteoric matter (*Nature*, 28–105). *See Nature*, 27–315, 338, 365, 388, 412, 434.

"Very little doubt it was an electric phenomenon" (Proctor, *Knowledge*, 2–419).

In the London *Times*, Nov. 20, 1882, the Editor says that he had received a great number of letters upon this phenomenon. He publishes two. One correspondent describes it as "well-defined and shaped like a fish ... extraordinary and alarming". The other correspondent writes of it as "a most magnificent luminous mass, shaped somewhat like a torpedo".

304

Chapter 26

Notes and Queries, 5–3–306:

About eight lights that were seen in Wales, over an area of about eight miles, all keeping their own ground, whether moving together perpendicularly, horizontally, or over a zigzag course. They looked like electric lights—disappearing, reappearing dimly then shining as bright as ever. "We have seen them three or four at a time afterward, on four or five occasions."

London *Times*, Oct. 5, 1887:

"From time to time the west coast of Wales seems to have been the scene of mysterious lights...." And now we have a statement from Towyn that within the last few weeks lights of various colours have been seen moving over the estuary of the Dysynni River, and out to sea. They are generally in a northerly direction, but sometimes they hug the shore, and move at high velocity for miles toward Aberdovey, and suddenly disappear.

L'Année Scientifique, 1877–45:

Lights that appeared in the sky, above Vence, France, March 23, 1877; described as balls of fire of dazzling brightness; appeared from a cloud about a degree in diameter; moved relatively slowly. They were visible more than an hour, moving northward. It is said that eight or ten years before similar lights or objects had been seen in the sky, at Vence.

London *Times*, Sept. 19, 1848:

That, at Inverness, Scotland, two large, bright lights that looked like stars had been seen in the sky: sometimes stationary, but occasionally moving at high velocity.

L'Année Scientifique, 1888–66:

Observed near St. Petersburg, July 30, 1880, in the evening: a large spherical light and two smaller ones, moving along a ravine; visible three minutes; disappearing without noise.

Nature, 35–173:

That, at Yloilo, Sept. 30, 1886, was seen a luminous object the size of the full moon. It "floated" slowly "northward", followed by smaller ones close to it.

"The False Lights of Durham."

Every now and then in the English newspapers, in the middle of the nineteenth century, there is something about lights that were seen against the sky, but as if not far above land, oftenest upon the coast of Durham. They were mistaken for beacons by sailors. Wreck after wreck occurred. The fishermen were accused of displaying false lights and profiting by wreckage. The fisherman answered that mostly only old vessels, worthless except for insurance, were so wrecked.

In 1866 (London *Times*, Jan. 9, 1866) popular excitement became intense. There was an investigation. Before a commission, headed by Admiral Collinson, testimony was taken. One witness described the light that had deceived him as "considerably elevated above ground". No conclusion was reached: the lights were called "the mysterious lights". But whatever the "false lights of Durham" may have been, they were unaffected by the investigation. In 1867, the Tyne Pilotage took the matter up. Opinion of the Mayor of Tyne—"a mysterious affair".

Jour. Roy. Astro. Soc. of Canada, November and December, 1913.

That, according to many observations collected by Prof. Chard of Toronto, there appeared, upon the night of Feb. 9, 1913, a spectacle that was seen in Canada, the United States, and at sea, and in Bermuda. A luminous body was seen. To it there was a long tail. The body grew rapidly larger. "Observers differ as to whether the body was single, or was composed of three or four parts, with a tail to each part." The group, or complex structure, moved with "a peculiar, majestic deliberation". "It disappeared in the distance, and another group emerged from its place of origin. Onward they moved, at the same deliberate pace, in twos or threes or fours." They disappeared. A third group, or a third structure followed.

Some observers compared the spectacle to a fleet of airships: others to battleships attended by cruisers and destroyers.

According to one writer:

"There were probably 30 or 32 bodies, and the peculiar thing about them was their moving in fours and threes and twos abreast of one another; and so perfect was the lining

306

up that you would have thought it was an aerial fleet manoeuvring after rigid drilling."

Nature, May 25, 1893:

A letter from Capt. Charles J. Norcock, of H.M.S. *Caroline*:

That, upon the 24th of February, 1893, at 10 P.M., between Shanghai and Japan, the officer of the watch had reported "some unusual lights".

They were between the ship and a mountain. The mountain was about 6,000 feet high. The lights seemed to be globular. They moved sometimes massed, but sometimes strung out in an irregular line. They bore "northward", until lost to sight. Duration two hours.

The next night the lights were seen again. They were, for the time, eclipsed by a small island. They bore north at about the same speed and in about the same direction as speed and direction of the *Caroline*. But they were lights that cast a reflection: there was a glare upon the horizon under them. A telescope brought out but few details: that they were reddish, and seemed to emit a faint smoke. This time the duration was seven and a half hours.

Then Capt. Norcock says that, in the same general locality, and at about the same time, Capt. Castle, of H.M.S. *Leander*, had seen lights. He had altered his course and had made toward them. The lights had fled from him. At least, they had moved higher in the sky,

Monthly Weather Review, March, 1904–115:

Report from the observations of three members of his crew by Lieut. Frank H. Schofield, U.S.N., of the U.S.S. *Supply*:

Feb. 24, 1904. Three luminous objects, of different sizes, the largest having an apparent area of about six suns. When first sighted, they were not very high. They were below clouds of an estimated height of about one mile.

They fled, or they evaded, or they turned.

They went up into the clouds which they had, at first, been sighted.

Their unison of movement.

But they were of different sizes, and of different susceptibilities, to all forces of this earth and of the air.

Monthly Weather Review, August, 1898–358:

Two letters from C. N. Crotsenburg, Crow Agency, Montana:

That, in the summer of 1896, when this writer was a railroad postal clerk—or one who was experienced in train-phenomena—while his train was going "northward", from Trenton, Mo., he and another clerk saw, in the darkness of a heavy rain, a light that appeared to be round, and of a dull-rose colour, and seemed to be about a foot in diameter. It seemed to float within a hundred feet of the earth, but soon rose high, or "midway between horizon and zenith". The wind was quite strong from the east, but the light held a course due north.

Its speed varied. Sometimes it seemed to outrun the train "considerably". At other times it seemed to fall behind. The mail clerks watched until the town of Linville, Iowa, was reached. Behind the depot of this town, the light disappeared, and was not seen again. All this time there had been rain, but very little lightning, but Mr. Crotsenburg offers the explanation that it was "ball lightning".

The Editor of the *Review* disagrees. He thinks that the light may have been a reflection from the rain, or fog, or from leaves of trees, glistening with rain, or the train's light—not lights.

In the December number of the *Review* is a letter from Edward M. Boggs—that the light was a reflection, perhaps, from the glare—one light, this time—from the locomotive's fire-box, upon wet telegraph wires—an appearance that might not be striated by the wires, but consolidated into one rotundity—that it had seemed to oscillate with the undulations of the wires, and had seemed to change horizontal distance with the varying angles of reflection, and had seemed to advance or fall behind, when the train had rounded curves.

All of which is typical of the best quasi-reasoning. It includes and assimilates diverse data: but it excludes that which will destroy it:

That, acceptably, the telegraph wires were alongside the track beyond, as well as leading to Linville.

Mr. Crotsenburg thinks of "ball lightning", which, though a sore bewilderment to most speculation, is usually supposed to be a correlate with the old system of thought: but his awareness of "something else" is expressed in other parts of his letters, when he says that he has something to tell that is "so strange that I should never have mentioned it, even to my friends, had it not been corroborated ... so unreal that I hesitated to speak of it, fearing that it was some freak of the imagination".

Chapter 27

Vast and black. The thing that was poised, like a crow over the moon.

Round and smooth. Cannon balls. Things that have fallen from the sky to this earth.

Our slippery brains.

Things like cannon balls have fallen, in storms, upon this earth. Like cannon balls are things that, in storms, have fallen to this earth.

Showers of blood.

Showers of blood.

Showers of blood.

Whatever it may have been, something like red-brick dust, or a red substance in a dried state, fell at Piedmont, Italy, Oct. 27, 1814 (*Electric Magazine*, 68–437). A red powder fell, in Switzerland, winter of 1867 (*Pop. Sci. Rev.*, 10–112)—

That something, far from this earth, had bled—super-dragon that had rammed a comet—

Or that there are oceans of blood somewhere in the sky—substance that dries, and falls in a powder—wafts for ages in powdered form—that there is a vast area that will some day be known to aviators as the Desert of Blood. We attempt little of super-topography, at present, but Ocean of Blood, or Desert of Blood—or both—Italy is nearest to it—or to them.

I suspect that there were corpuscles in the substance that fell in Switzerland, but all that could be published in 1867 was that in this substance there was a high proportion of "variously shaped organic matter".

At Giessen, Germany, in 1821, according to the *Report of the British Association*, 5–2, fell a rain of a peach-red colour. In this rain were flakes of a hyacinthine tint. It is said that this substance was organic: we are told that it was pyrrhine.

But distinctly enough, we are told of one red rain that it was of corpuscular composition—red snow, rather. It fell, March 12, 1876, near the Crystal Palace, London (*Year Book of Facts*, 1867–89; *Nature*, 13–414). As to the "red snow" of polar and mountainous regions, we have no opposition, because that "snow" has never been seen to fall from the sky: it is growth of micro-organisms, or of a "protococcus," that spreads over snow that is on the ground. This time nothing is said of "sand from the Sahara". It is said of the red matter that fell in London, March 2, 1876, that it was composed of corpuscles—

Of course:

That they looked like "vegetable cells".

A note:

That nine days before had fallen the red substance—flesh—whatever it may have been—of Bath County, Kentucky.

I think that a super-egotist, vast, but not so vast as it had supposed, had refused to move to one side for a comet.

We summarize our general super-geographical expressions:

Gelatinous regions, sulphurous regions, frigid and tropical regions: a region that has been Source of Life relatively to this earth: regions wherein there is density so great that things from them, entering this earth's thin atmosphere, explode.

We have had a datum of explosive hailstones. We now have support to the acceptance that they had been formed in a medium far denser than air of this earth at sea-level. In the *Popular Science News*, 22–38, is an account of ice that had been formed, under great pressure, in the laboratory of the University of Virginia. When released and brought into contact with ordinary air, this ice exploded.

And again the flesh-like substance that fell in Kentucky: its flake-like formation. Here is a phenomenon that is familiar to us: it suggests flattening, under pressure. But the extraordinary inference is—pressure not equal on all sides. In the *Annual Record of Science*, 1873–350, it is said that, in 1873, after a heavy thunderstorm in Louisiana, a tremendous number of fish scales were found, for a distance of forty miles, along the banks of the Mississippi River: bushels of them picked up in single places: large scales that were said to be of the gar fish, a fish that weighs from five to fifty pounds. It seems impossible to accept this identification: one thinks of a substance that had been pressed into flakes or scales. And round hailstones with wide thin margins of ice irregularly around them—still such hailstones seem to me more like things that had been stationary: had been held in a field of thin ice. In the *Illustrated London News*, 34–546, are drawings of hailstones so margined, as if they had been held in a sheet of ice.

Some day we shall have an expression which will be, to our advanced primitiveness, a great joy:

That devils have visited this earth: foreign devils: human-like beings, with pointed beards: good singers; one shoe ill-fitting—but with sulphurous exhalations, at any rate. I have been impressed with the frequent occurrence of sulphurousness with things that come from the sky. A fall of jagged pieces of ice, Orkney, July 24, 1818 (*Trans. Roy. Soc. Edin*, 9–187). They had a strong sulphurous odour. And the coke—or the substance that looked like coke—that fell at Mortrée, France, April 24, 1887: with it fell a sulphurous substance. The enormous round things that rose from the ocean, near the *Victoria*. Whether we still accept that they were super-constructions that had come from a denser atmosphere and, in danger of disruption, had plunged into the ocean for relief, then rising and continuing on their way to Jupiter or Uranus—it was reported that they spread a "stench of sulphur". At any rate, this datum of proximity is against the conventional explanation that these things did not rise from the ocean, but rose far away above the horizon, with illusion of nearness.

And the things that were seen in the sky July, 1898: I have another note. In *Nature*, 58–224, a correspondent writes that, upon July 1, 1898, at Sedberg, he had seen in the sky—a red object—or, in his own wording, something that looked like the red part of a rainbow, about 10 degrees long. But the sky was dark at the time. The sun had set. A heavy rain was falling.

Throughout this book, the datum that we are most impressed with:

Successive falls.

Or that, if upon one small area, things fall from the sky, and then, later, fall again upon the same small area, they are not products of a whirlwind which though sometimes axially stationary, discharges tangentially—

So the frogs that fell at Wigan. I have looked that matter up again. Later more frogs fell.

As to our data of gelatinous substance said to have fallen to this earth with meteorites, it is our expression that

311

meteorites, tearing through the shaky, protoplasmic seas of Genesistrine—against which we warn aviators, or they may find themselves suffocating in a reservoir of life, or struck like currants in a blancmange—that meteorites detach gelatinous, or protoplasmic, lumps that fall with them.

Now the element of positiveness in our composition yearns for the appearance of completeness. Super-geographical lakes with fishes in them. Meteorites that plunge through these lakes, on their way to this earth. The positiveness in our make-up must have expression in at least one record of a meteorite that has brought down a lot of fishes with it—

Nature, 3–512:

That, near the bank of a river, in Peru, Feb. 4, 1871, a meteorite fell. "On the spot, it is reported, several dead fishes were found, of different species." The attempt to correlate is—that the fishes "are supposed to have been lifted out of the river and dashed against the stones".

Whether this be imaginable or not depends upon each one's own hypnoses.

Nature, 4–169:

That the fishes had fallen among the fragments of the meteorite.

Popular Science Review, 4–126:

That one day, Mr. Le Gould, an Australian scientist, was travelling in Queensland. He saw a tree that had been broken off close to the ground. Where the tree had been broken was a great bruise. Near by was an object that "resembled a ten-inch shot".

A good many pages back there was an instance of overshadowing, I think. The little carved stone that fell at Tarbes is my choice as the most impressive of our new correlates. It was coated with ice, remember. Suppose we should sift and sift and discard half the data in this book—suppose only that one datum should survive. To call attention to the stone of Tarbes would, in my opinion, be doing well enough, for whatever the spirit of this book is trying to do. Nevertheless, it seems to me that a datum that preceded it was slightingly treated.

The disc of quartz, said to have fallen from the sky, after a meteoric explosion:

Said to have fallen at the plantation Bleijendal, Dutch Guinea: sent to the Museum of Leyden by M. van Sypesteyn, adjutant to the Governor of Dutch Guiana (*Notes and Queries*, 2–8–92).

And the fragments that fall from super-geographic ice fields: flat pieces with icicles on them. I think that we did not emphasize enough that, if these structures were not icicles, but crystalline protuberances, such crystalline formulations indicate long suspension quite as notably as would icicles. In the *Popular Science News*, 24–34, it is said that in 1869, near Tiflis, fell large hailstones with long protuberances. "The most remarkable point in connection with the hailstones is the fact that, judging from our present knowledge, a very long time must have been occupied in their formation." According to the *Geological Magazine*, 7–27, this fall occurred May 27, 1869. The writer in the *Geological Magazine* says that of all theories that he had ever heard of, not one could give him light as to this occurrence—"these growing crystalline forms must have been suspended a long time"—

Again and again this phenomenon:

Fourteen days later, at about the same place, more of these hailstones fell.

Rivers of blood that vein albuminous seas, or an egg-like composition in the incubation of which this earth is a local centre of development—that there are super-arteries of blood in Genesistrine: that sunsets are consciousness of them: that they flush the skies with northern lights sometimes: super-embryonic reservoirs from which life-forms emanate—

Or that our whole solar system is a living thing: that showers of blood upon this earth are its internal haemorrhages—

Or vast living things in the sky, as there are vast living things in the oceans—

Or some one especial thing: an especial time: an especial place. A thing the size of the Brooklyn Bridge. It's alive in outer space—something the size of Central Park kills it—

It drips.

We think of the ice fields above this earth: which do not, themselves, fall to this earth, but from which water does fall—

Popular Science News, 35–104:

That according to Prof. Luigi Palazzo, head of the Italian Meteorological Bureau, upon May 15, 1890, at

Messignadia, Calabria, something of the colour of fresh blood fell from the sky.

This substance was examined in the public-health laboratories of Rome.

It was found to be blood.

"The most probable explanation of this terrifying phenomenon is that migratory birds (quails or swallows) were caught and torn in a violent wind."

So the substance was identified as birds' blood—

What matters is that the microscopes of Rome said—or had to say—and what matters is that we point out that here is no assertion that there was a violent wind at the time—and such a substance would be almost infinitely dispersed in a violent wind—that no bird was said to have fallen from the sky—or said to have been seen in the sky—that not a feather of a bird is said to have been seen—

This old datum:

The fall of blood from the sky—

But later, in the same place, blood again fell from the sky.

Chapter 28

Notes and Queries, 7–8–508:

A correspondent who had been to Devonshire writes for information as to a story that he had heard there: of an occurrence of about thirty-five years before the date of writing:

Of snow upon the ground—of all South Devonshire waking up one morning to find such tracks in the snow as had never before been heard of—"clawed footmarks" or "an unclassifiable form"—alternating at huge but regular intervals with what seemed to be the impression of the point of a stick—but the scattering of the prints—amazing expanse of territory covered—obstacles, such as hedges, walls, houses, seemingly surmounted—

Intense excitement—that the track had been followed by huntsmen and hounds, until they had come to a forest—from which the hounds had retreated, baying and terrified, so that no one had dared to enter the forest.

Notes and Queries, 7–9–18:

Whole occurrence well-remembered by a correspondent: a badger had left marks in the snow: this was determined, and the excitement had "dropped to a dead calm in a single day".

Notes and Queries, 7–9–70:

That for years a correspondent had had a tracing of the prints, which his mother had taken from those in the snow in her garden, in Exmouth: that they were hoof-like marks—but had been made by a biped.

Notes and Queries, 7–79–253:

Well remembered by another correspondent, who writes of the excitement and consternation of "some classes". He says that a kangaroo had escaped from a menagerie—"the footprints being so peculiar and far apart gave rise to a scare that the devil was loose".

We have had a story, and now we shall tell it over from contemporaneous sources. We have had the later accounts first very largely for an impression of the correlating effect that time brings about, by addition, disregard and distortion.

For instance, the "dead calm in a single day". If I had found that the excitement did die out rather soon, I'd incline to accept that nothing extraordinary had occurred.

I found that the excitement had continued for weeks.

I recognize this as a well-adapted thing to say, to divert attention from a discorrelate.

All phenomena are "explained" in the terms of the Dominant of their era. This is why we give up trying really to explain, and content ourselves with expressing. Devils that might print marks in snow are correlates to the third Dominant back from this era. So it was an adjustment by nineteenth-century correlates, or human tropisms, to say that the marks in the snow were clawed. Hoof-like marks are not only horsy but devilish. It had to be said in the nineteenth century that those prints showed claw-marks. We shall see that this was stated by Prof. Owen, one of the greatest biologists of his day—except that Darwin didn't think so. But I shall give reference to two representations of them that can be seen in the New York Public Library. In neither representation is there the faintest suggestion of a claw-mark. There never has been a Prof. Owen who has explained: he has correlated.

Another adaptation, in the later accounts, is that of leading this discorrelate to the Old Dominant into the familiar scenery of a fairy story, and discredit it by assimilation to the conventionally fictitious—so the idea of the baying, terrified hounds, and forest like enchanted forests, which no one dared to enter. Hunting parties were organized, but the baying, terrified hounds do not appear in contemporaneous accounts.

The story of the kangaroo looks like adaptation to needs for an animal that could spring far, because marks were found in the snow on roof of houses. But so astonishing is the extent of snow that was marked that after a while another kangaroo was added.

But the marks were in single lines.

My own acceptance is that not less than a thousand one-legged kangaroos, each shod with a very small horseshoe, could have marked that snow of Devonshire.

London *Times*, Feb. 16, 1855:

"Considerable sensation has been caused in the towns of Topsham, Lymphstone, Exmouth, Teignmouth, and Dawlish in Devonshire, in consequence of the discovery of a vast number of foot tracks of a most strange and mysterious description."

The story is of an incredible multiplicity of marks discovered in the morning of Feb. 8, 1855, in the snow, by the inhabitants of many towns and regions between towns. This great area must of course be disregarded by Prof. Owen and the other correlators. The tracks were in all kinds of unaccountable places: in gardens enclosed by high walls, and up on the tops of houses, as well as in the open fields. There was in Lymphstone scarcely one unmarked garden. We've had heroic disregards but I think that here disregard was titanic. And, because they occurred in single lines, the marks are said to have been "more like those of a biped than of a quadruped"—as if a biped would place one foot precisely ahead of another—unless it hopped—but then we have to think of a thousand, or of thousands.

It is said that the marks were "generally eight inches in advance of each other".

"The impression of the foot closely resembles that of a donkey's shoe, and measured from an inch and a half, in some instances, to two and a half inches across."

Or the impressions were cones in incomplete, or crescentic basins.

The diameters equalled diameters of very young colts' hoofs: too small to be compared with marks of donkeys' hoofs.

"On Sunday last the Rev. Mr. Musgrave alluded to the subject in his sermon and suggested the possibility of the footprints being those of a kangaroo, but this could scarcely have been the case, as they were found on both sides of the Este. At present it remains a mystery, and many superstitious people in the above-named towns are actually afraid to go outside their doors after night."

The Este is a body of water two miles wide.

London, *Times*, March 6, 1855:

"The interest in this matter has scarcely yet subsided, many inquiries still being made into the origin of the footprints, which caused so much consternation upon the morning of the 8th ult. In addition to the circumstances mentioned in *The Times* a little while ago, it may be stated that at Dawlish a number of persons sallied out, armed with guns and other weapons, for the pupose, if possible, of discovering and destroying the animal which was supposed

317

to have been so busy in multiplying its footprints. As might have been expected, the party returned as they went. Various speculations have been made as to the cause of the footprints. Some have asserted that they are those of a kangaroo, while others affirm that they are the impressions of claws of large birds driven ashore by stress of weather. On more than one occasion reports have been circulated that an animal from a menagerie had been caught, but the matter at present is as much involved in mystery as ever it was."

In the *Illustrated London News*, the occurrence is given a great deal of space. In the issue of Feb. 24, 1855, a sketch is given of the prints.

I call them cones in incomplete basins.

Except that they're a little longish, they look like the prints of hoofs of horses—or, rather, of colts.

But they're in a single line.

It is said that the marks from which the sketch was made were eight inches apart, and that this spacing was regular and invariable "in every parish". Also other towns besides those named in *The Times* are mentioned. The writer, who had spent a winter in Canada, and was familiar with tracks in snow, says that he had never seen "a more clearly defined track". Also he brings out the point that was so persistently disregarded by Prof. Owen and the other correlators—that "no known animal walks in a line of single footsteps, not even man". With these wider inclusions, this writer concludes with us that the marks were not footprints. It may be that his following observation hits upon the crux of the whole occurrence:

That whatever it may have been that had made the marks, it had removed, rather than pressed, the snow.

According to his observations the snow looked "as if branded with a hot iron".

Illustrated London News, March 3, 1855–214:

Prof. Owen, to whom a friend had sent drawings of the prints, writes that there were claw-marks. He says that the "track" was made by "a" badger.

Six other witnesses sent letters to this number of the *News*. One mentioned, but not published, is a notion of a strayed swan. Always this homogeneous-seeing—"a" badger—"a" swan—"a" track. I should have listed the other towns as well as those mentioned in *The Times*.

A letter from Mr. Musgrave is published. He, too, sends a sketch of the prints. It, too, shows a single line. There are four prints, of which the third is a little out of line.

There is no sign of a claw-mark.

The prints look like prints of longish hoofs of a very young colt, but they are not so definitely outlined as in the sketch of February 24th, as if drawn after disturbance by wind, or after thawing had set in. Measurements at places a mile and a half apart, gave the same interspacing—"exactly eight inches and a half apart".

We now have a little study in the psychology and genesis of an attempted correlation. Mr. Musgrave says: "I found a very apt opportunity to mention the name 'kangaroo' in allusion to the report then current." He says that he had no faith in the kangaroo-story himself, but was glad "that a kangaroo was in the wind", because it opposed "a dangerous, degrading, and false impression that it was the devil".

"Mine was a word in season and did good."

Whether it's Jesuitical or not, and no matter what it is or isn't, that is our own acceptance: that, though we've often been carried away from this attitude controversially, that is our acceptance as to every correlate of the past that has been considered in this book—relatively to the Dominant of its era.

Another correspondent writes that, though the prints in all cases resembled hoof marks, there were indistinct traces of claws—that "an" otter had made the marks. After that many other witnesses wrote to the *News*. The correspondence was so great that, in the issue of March 10th, only a selection could be given. There's "a" jumping-rat solution and "a" hopping-toad inspiration, and then someone came out strong with an idea of "a" hare that had galloped with pairs of feet held close together, so as to make impressions in a single line.

London *Times*, March 14, 1840:

"Among the high mountains of that elevated district where Glenorchy, Glenlyon and Glenochay are contiguous, there have been met with several times, during this and also the former winter, upon the snow, the tracks of an animal seemingly unknown at present in Scotland. The print, in every respect, is an exact resemblance to that of a foal of considerable size, with this small difference, perhaps, that the sole seems a little longer or not so round;

but as no one has had the good fortune as yet to have obtained a glimpse of this creature, nothing more can be said of its shape or dimensions; only it has been remarked, from the depth to which the feet sank in the snow, that it must be a beast of considerable size. It has been observed also that its walk is not like that of the generality of quadrupeds, but that it is more like the bounding or leaping of a horse when scared or pursued. It is not in one locality that its tracks have been met with, but through a range of at least twelve miles."

In the *Illustrated London News*, March 17, 1855, a correspondent from Heidelberg writes, "upon the authority of a Polish Doctor of Medicine", that on the Piashowa-gora (Sand Hill) a small elevation on the border of Galicia, but in Russian Poland, such marks are to be seen in the snow every year, and sometimes in the sand of this hill, and "are attributed by the inhabitants to supernatural influences".